NANOTECHNOLOGY SCIENCE AND TECHNOLOGY

TRIBOMECHANICAL MODIFICATION OF FRICTION SURFACE BY RUNNING-IN IN LUBRICANTS WITH NANO-SIZED DIAMONDS

NANOTECHNOLOGY SCIENCE AND TECHNOLOGY

NANOTECHNOLOGY SCIENCE AND TECHNOLOGY

TRIBOMECHANICAL MODIFICATION OF FRICTION SURFACE BY RUNNING-IN IN LUBRICANTS WITH NANO-SIZED DIAMONDS

P.A. VITYAZ
V.I. ZHORNIK
V.A. KUKAREKO
AND
M.A. BELOTSERKOVSKY

Nova Science Publishers, Inc.
New York

For permission to use material from this book please contact us:
Telephone 631-231-7269; Fax 631-231-8175
Web Site: http://www.novapublishers.com

Library of Congress Cataloging-in-Publication Data

Available upon Request
ISBN: 978-1-61728-921-7

Published by Nova Science Publishers, Inc. † New York

CONTENTS

PREFACE

The book presents a brief analysis of such methods of tribotechnology as the formation of coatings from a filler material on part surfaces owing to friction forces (the frictional facing of wear-resistant coatings and finish antifriction abrasive-free treatment) and coating deposition during friction with a lubricant containing plating components (metal-plating compositions and cermets).

Peculiarities of such method of tribotechnology as running-in in a lubricant containing nano-sized hard components, which include nanodiamonds are considered in detail. It is shown that tribomechanical modification in the presence of lubricants with nanodiamonds improves tribological properties of friction pairs and prolongs the life of friction units. Hard nanoparticles introduced into a lubricant have a complex effect on tribosystems increasing their durability. Nanoparticles improve properties of lubricants and modify the contacting surfaces. Among the processes affecting positively a tribosystem are the following: the formation of separating layers with a higher carrying capacity, the stabilization of rheological characteristics of a lubricating film within a broad temperature range, the smoothening of friction surfaces, and the hardening of surface layers.

Nanoparticles introduced into a grease act as centers of complexation and favor the formation of a more branched structure frame of the grease disperse phase, which provides a high oil-retaining capacity and widens the working load range. The efficiency of the nano-sized diamond-containing additive is shown using lithium grease and industrial oil as examples.

The influence of the nano-sized hard additive to lubricants on the tribological behavior of friction units is discussed for friction pairs made of metals and metal-polymer composites. During friction, severe plastic deformation caused by the effect of hard nanoparticles accelerates the running-in of the friction surfaces. Additionally, the surface microhardness increases (up to 2 times and more) and the nano-sized substructure with a subgrain size of < 100 nm appears in the surface layer of ductile materials. It possesses a high resistance to the nucleation and propagation of fatigue microcracks at friction. The accelerated formation of hardened surface layers and the intensive running-in of friction surfaces prevent them from adhesion under heavy pressures. A correlation has been found between the modifying effect and the load and the initial hardness of a material. A wear model of a friction surface in the presence of a lubricant containing hard nano-sized particles has been proposed. It is shown that the positive effect of the nano-sized diamond-containing component becomes stronger with increasing the contact load and decreasing the initial material hardness. The application of lubricants with the nano-sized diamond-containing additives reduces the friction coefficient 1.5–2 times and increases the wear resistance up to 30 times.

It has been found that the formation of the wear-resistant structure in the surface layer occurs during the initial stages of friction unit operation, i.e. during running-in. Subsequently, common lubricants can be used, but high antifriction properties and wear resistance of the unit retain.

The book presents results of the study of structural-phase transformations that occur during friction in the surface layer of gas-thermal coatings from steels of austenite and martensite classes deposited by various methods (gas-flame spraying and electric arc metallization). During lubrication with a lubricant containing hard nano-sized additives the $\gamma \to \alpha$ phase transformation of metastable austenite to martensite evolves in metastable austenite steels. In this case the content of the martensite α-phase in the surface layer increases up to 35–40 vol.% and the severe hardening of the layer occurs. The scoring resistance of the coatings having the metastable structure augments by 3–4 times and the carrying capacity of coatings from, for example, steel 12Kh18N10T reaches 100 MPa.

The authors discuss the potential of applying the method of the tribomechanical modification of friction surfaces by running-in in a lubricant with nanodiamonds to friction units of various machines and mechanisms such as quarry and agricultural machinery, heat-and-power plants, etc.

INTRODUCTION

Nowadays various methods of the treatment of surfaces of friction members are widely used to give them required tribological properties. Most of them involve coating deposition on a friction surface or the modification of the structure of a surface layer using electrophysical, electrochemical, thermochemical and other modes of effect on a material.

Tribotechnology is a new promising direction of technology aimed at the improvement of tribological properties of friction surfaces and the prolongation of the life of friction units. It is based on the principle of the transformation of the destructive effect of the friction force to the creative process of the hardening and recovery of friction surfaces as well as on the principle of controlling physical-chemical transformations occurring on a friction surface owing to thermomechanical effect in friction contact. The aim of tribotechnology is to form structures with a lower shear energy or a higher resistance to fatigue crack nucleation in a surface layer as well as to recover the worn surface layer.

Tribotechnology includes coating deposition by frictional facing when the material being deposited is melted by the heat liberated in friction of the filler material against the part surface being covered.

Among the most famous phenomena used in tribotechnology is selective transfer. Selective transfer in friction or the so-called effect of wearless friction results from the occurrence of chemical and physical processes on the contacting surfaces which lead to the formation of self-organizing systems of self-compensation of the wear and reduction of the friction coefficient. Selective transfer is a mode of internal friction in a spontaneously appeared metallic (servovite) film which is formed on friction surfaces at a definite combination of the contacting materials and the lubricant. A special deformation mechanism occurs in the servovite film which is not accompanied by the accumulation of defects typical for fatigue processes. However, the practical application of selective transfer if often problematic since it is difficult to keep the conditions of its realization.

The principles of selective transfer form a basis for methods of finish antifriction abrasive-free treatment of friction surfaces. It includes the deposition of a thin layer of a ductile material on the treated surface by rubbing; the material possesses improved antifriction properties.

A wide class of metal-plating lubricants is known which contain ultradispersed powders of soft metals or their chemical compounds. Their principle of operation involves the formation of protective films on real contact spots. The films localize shear deformations in friction and reduce the force interaction and wear rate of the contacting surfaces.

Tribotechnology includes methods of recovering friction surfaces during their operation using additives to lubricants which contain dispersed natural layered silicates. The application of such additives favors the recover of worn surfaces by forming cermet coatings on them having an improved wear resistance.

Techniques for improving tribological properties of friction pairs and prolonging the life of friction units by their running-in with a lubricant containing nano-sized hard components, including nano-diamonds, are also considered as a direction of tribotechnology. Hard nano-sized particles introduced to a lubricant exert complex influence on a tribosystem thus prolonging its durability. From the one hand, nanoparticles improve physical and mechanical properties of the lubricant and increase the carrying capacity of the lubricating film. From the second hand, hard nano-sized particles are capable of affecting positively the contacting surfaces by varying their structure in friction. Among the processes affecting positively a tribosystem when using lubricants with nano-sized hard additives are the formation of separating layers with an improved carrying capacity, the stabilization of rheological characteristics of a lubricating film within a broad temperature range, decrease in the friction surface roughness and the hardening of surface layers.

The application of lubricants modified by hard nano-sized additives intensifies the running-in of friction units, widens the range of temperature and load regimes of operation of friction units and prolongs their life.

Chapter 1

METHODS OF TRIBOMECHANICAL TREATMENT OF SURFACES

1.1. TRIBOMECHANICAL FORMATION OF COATINGS FROM FILLER MATERIAL

1.1.1. Formation of Thick Coatings by Frictional Facing

The deposition of coatings from granulated materials by frictional facing is characterized by the presence of the following two physical components: thermal and mechanical. The role of each component can be selected conventionally in the following way. The thermal factor causes the severe heating of the material being deposited and the mated base metal to the melting temperature of the former; it favors the occurrence of diffusion processes. The mechanical factor governs the formation of pore-free structures.

The development of a method for coating deposition using the frictional energy included the following stages:

- finding a relationship between frictional facing and the known technologies of the hardening and treatment of metals involving the joint effect of elevated temperatures and mechanical loads;
- the study of the influence of temperature-time conditions and the stress state of the zone of contact between the deposited material and base surface on the mechanism of the formation of secondary

structures in the contact zone at frictional facing of alloys having various metallophysical nature;
- the study of the influence of technological parameters of frictional facing on the durability of the bimetallic articles fabricated;
- the determination of rational technological regimes of the deposition of wear-resistant coatings by frictional facing and the analysis of fields of its effective application.

The technological realization of frictional facing needs no complicated devices. The scheme of a setup for coating deposition by frictional facing depends much on the shape of parts which are divided into the following two groups: parts like a shaft with the concentric disposition of all the surfaces relatively to one axis and parts whose center is shifted from the axis of the cylindrical surface being coated. Figure 1.1 shows the schematic diagram of a device for covering the parts of the fist type. It serves to produce contact between a monolytic base (a shaft or rod) and a granulated (powdered or chips-like) briquetted coating material. The monolytic base takes the torque that, in combination with the motion of the forming tool (cams), provides the development of friction regimes (the relative sliding and pressure) in the zone of friction contact between the steel base surface and the granulated material. This, in its turn, causes severe heat generation sufficient for the bulk of the deposited alloy to melt.

Despite the seeming kinematic simplicity of the devises used, friction facing is accompanied by quite complex physical-mechanical and thermal phenomena. Severe friction regimes in the zone of contact between the granulated briquette and shaft cause intensive heat generation; the selection of technological parameters provides one of the necessary conditions, i.e. the melting of the deposited granulated alloy. Under the effect of forces of capillary pressure the bulk of the melted metal is transferred from the friction contact zone to the peripheral areas of the working chamber. As the forming tool moves, new parts of the briquette enter contact with the steel base surface until all the frame of the briquette becomes melted. The momentary stoppage of shaft rotation ceases heat generation and induces the development of melt crystallization with the joint effect of the external forming force.

Diffusion processes evolve actively in friction; they govern the strength of joint between the layers in the bimetal and the resistance of bimetallic articles to cyclic loading. Different thermal characteristics of the base and deposited layer in a fabricated bimetallic article cause the appearance of residual stresses which lead to significant variations in the performance characteristics of parts.

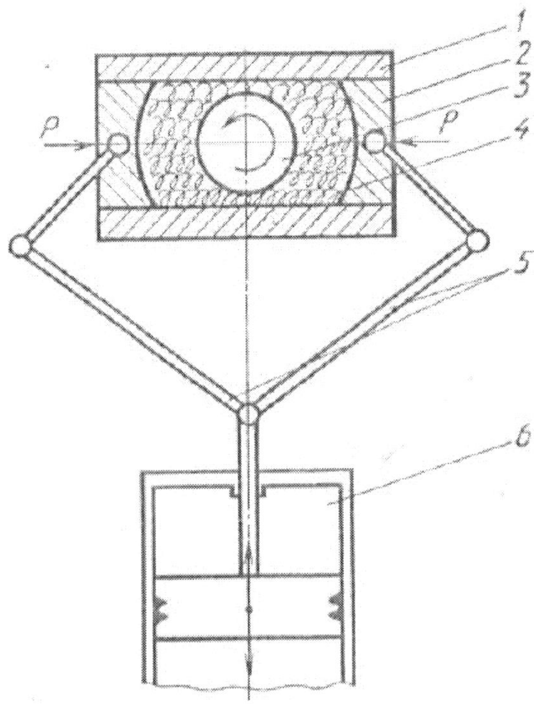

Figure 1.1. Schematic diagram of device for frictional facing of coatings on cylindrical surface [1]: *1* – stationary guide; *2* – shaping tool; *3* – steel base; *4* – deposited alloy in granulated and briquetted state; *5* – lever system; *6* – hydraulic cylinder.

Thus, we note the following phenomena accompanying the process of frictional facing: heat generation in the friction contact zone resulted from regularities of the friction of the granulated alloy against the steel base; diffusion processes caused by the friction of the alloys; the development of residual stresses owing to the temperature gradient through the section of an article occurring during its cooling and to different thermal characteristics of the deposited alloy and the steel base.

Kinematic peculiarities of frictional facing allow one to deposit coatings on a base whose shape belongs to any kind of a solid of revolution, primarily, cylindrical. Frictional facing is recommended for depositing fusible materials (copper and aluminum bearing alloys and, in some cases, wear-resistant alloys) having a relatively low melting temperature that is by 200 and more degrees lower than the melting temperature of the base material. Attempts of using frictional facing beyond the recommended domain give no positive results. Particularly, when depositing heterogeneous hard alloys, e.g. stellites, on a steel base, the alloy structure becomes coarser despite a relatively short process duration (80–90 s). This is caused by the initial period of the decomposition and coagulation of hard inclusions.

1.1.2. Finish Antifriction Abrasive-Free Treatment

Finish antifriction abrasive-free treatment is a method combining the deposition of a protective layer on a surface by friction and the surface hardening of the base metal. It is based on the use of the effect of selective transfer and favors considerable decrease in friction forces and the wear rate of the surfaces; besides, it requires little material, energy, and time consumption.

It was shown in papers [2–4] that in a steel–bronze pair lubricated with glycerol, under certain conditions including mixed and boundary lubrication, a state may appear that is characterized by exceptionally low friction forces comparable to those typical for fluid lubrication. The cause of this effect called selective transfer is the spontaneous appearance of a thin (1–2 μm thick) quasi-fluid metallic film in the contact zone during friction; it has a certain structure and specific tribological properties (figure 1.2).

The film is called a servovite film, i.e. self-regenerating. It is formed under the effect of the lubricant and loading conditions. The friction pair in which selective transfer is induced and kept is characterized by the following peculiarities: owing to selective dissolution the appeared servovite film has a defect (dislocation) structure (8–10% of lattice points remain free) that can give the film quasi-fluid properties. This favors decrease in friction forces and a considerable increase in the real contact area.

During selective transfer the Rehbinder effect or the effect of adsorption plasticization occurs because of the embedding of surfactant particles contained in the lubricant [5]. Surfactant molecules penetrate into pores of the servovite film thus promoting decrease in the shear strength in the friction zone. If a servovite film presents in the friction zone, wear particles contain mainly copper. They have a brittle and very active surface; for this reason they get covered with an adsorbed film of surfactants. Such particles called micelles carry an electric charge and under its effect they agglomerate into columns or bars. In addition, during selective transfer exchange with wear particles occurs which does not lead to the fracture of the friction surfaces. Since no oxide films are available at selective transfer, the servovite film acts as a catalyst of polymerization. A polymer film is formed from free radicals of the organic materials which appear during lubricant tribodestruction. This prevents the surfaces against metal–metal contact as well as reduces the maximal pressures and shear strength.

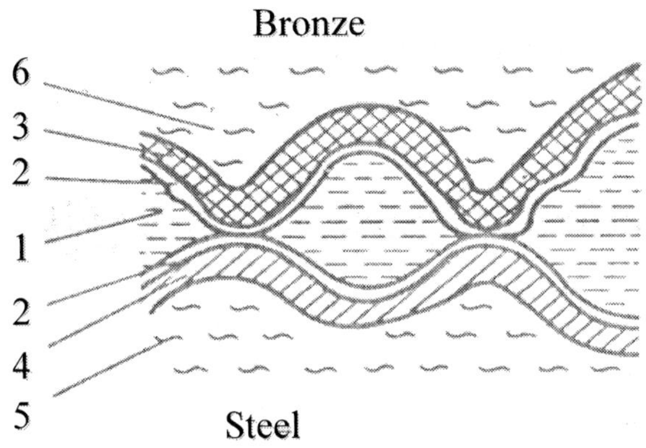

a

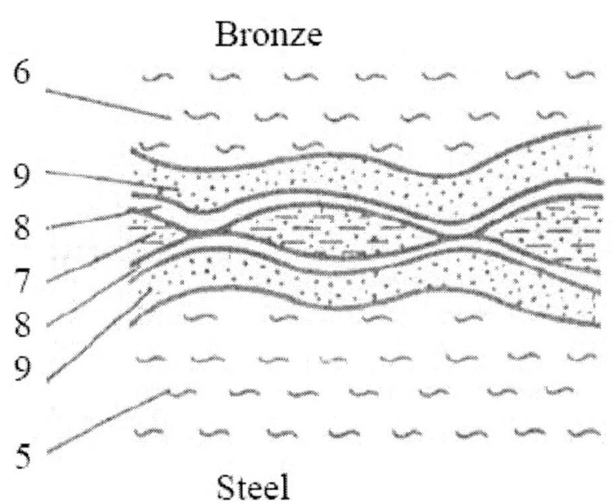

b

Figure 1.2. Reaction layers in bronze–steel friction pair lubricated with glycerol: *a* – original state; *b* – final state [4]; *1* – pure glycerol; *2* – adsorbed layers; *3* – copper layer 1–2 μm thick; *4* – iron oxide layer 1–2 μm thick; *5* –areas with steel lattice distortions; *6* – areas with bronze lattice distortions; *7* – glycerol with surfactant and wear products; *8* – adsorbed layers with surfactant and transformation products; *9* – copper layer with point defects and lubricant components.

Selective transfer or the effect of wearless friction is a mode of friction that is characterized by the action of mainly the molecular component of the friction force. It is the phenomenon opposite to wear since the latter leads to surface damage while selective transfer is of creative self-organizing nature. Selective transfer results from the occurrence of chemical and physical-chemical processes on a surface leading to the appearance of the systems of wear self-compensation and friction reduction.

A servovite film appears during selective transfer as a result of the decomposition of the copper alloy being a solid solution, particularly, without temperature elevation but in the presence of a lubricant containing surfactants which facilitate the diffusion process. The coalescence of pores in the servovite film occurs during diffusion and plastic deformation. In most cases pores are filled with the lubricant.

The formation of a servovite film in the bronze–steel contact with the lubricant providing the occurrence of selective transfer, e.g., glycerol is accompanied by relatively small material consumption. As soon as the film has appeared on the friction surfaces, the material consumption decreases. The film thickness depends on the copper alloy composition and peculiarities of the lubricant. For example, it is 23 μm for a brass–steel pair lubricated with glycerol and 0.5–1.0 μm for a tin bronze – steel pair. The consumption of nonferrous metals for the formation of the servovite film can be reduced by introducing a small amount of salt-forming copper-containing additives into the lubricant. Selective transfer can be realized in friction of pairs steel–bronze, steel–steel, steel – cast iron, cast iron – cast iron, steel – aluminum alloys and others. It may occur with various lubricants such as mineral and synthetic oils, greases, cooling lubricants etc [4].

The mechanical brass-plating of steel and cast iron friction surfaces is based on selective transfer.

Figure 1.3 shows the schematic of the mechanical brass-plating of a shaft. A rod made of a special brass alloy rotates around its axis and is pressed to a rotating surface being covered. Rubbing is performed under the certain velocity and pressure regimes and with commercially pure glycerol as a lubricant.

During mechanical brass plating, the surface hardening of a metal to a depth of up to 80 μm is possible in addition to the rubbing of a thin brass layer 2–4 μm thick adhering strongly to the base.

Carbon steels are easily covered in this way while alloyed steels and cast irons are more difficult to cover; the same is true for chrome-plated, phosphatized, and nickel-plated parts as well as for those made of aluminum alloys [5–7].

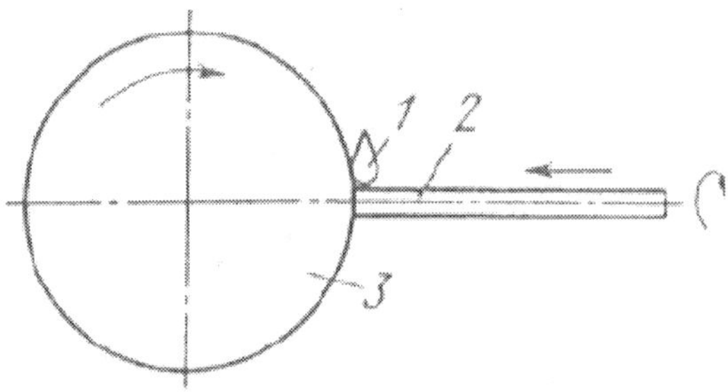

Figure 1.3. Schematic diagram of mechanical and chemical-mechanical brass-plating of shaft [4]: *1* – surfactant fluid (glycerol); *2* – rubbing brass rod; *3* – shaft.

Before deposition the surface is cleaned from oxide and other films which reduce or even eliminate microadhesion. The most important factors that influence the coating quality and rubbing conditions are as follows: the type of the material being coated and its state, the surface macrogeometry and conditions of heat removal, the roughness of the surface being coated, the material of the rubbing rod, the pressing force, sliding velocity, and the feed of the rubbing rod, the lubricant type, and the number of passes during rubbing.

During mechanical brass-plating the smoothing of surface layers occurs initially related to their activation under the effect of surfactant components of the lubricant. Microadhesion processes run simultaneously between the surface being coated and the rubbing brass rod. All these processes result in the formation of a transition layer on the steel or cast iron surface being covered and in the appearance of a brass coating.

Totally wearless friction can be achieved only in model friction pairs. In practice, a protective servovite film affects positively during a restricted period since no new copper particles enter the friction zone [5, 7].

In order to prevent the appearance of hairmarks on the surface the part material hardness should be no more than 30 HRC for steels and 150 HB for cast irons. The effective coating formation by finish antifriction abrasive-free treatment is provided when the surface roughness of a steel part is R_a = 0.63–2.50 μm. Cast iron surfaces can be rubbed mechanically only if their roughness is R_a = 5–10 μm [6].

The tool material should have the strength sufficient to break oxide films on the part surface in friction; it also should possess good ductility to make the contact area vaster. To provide the transfer of the tool material on the part material and eliminate back transfer, the tool strength should be considerably less than the part strength. In addition to special brass, other materials are used for rubbing that can reduce much the wear rate and friction force at mixed and boundary lubrication. It is noted in [3, 6] that copper-zinc alloys have the best technological properties while bronze and copper coatings are more difficult to deposit. The studies carried out in [8] have shown that the coatings of brasses with 35–41% of zinc have the best wear resistance (figure 1.4.).

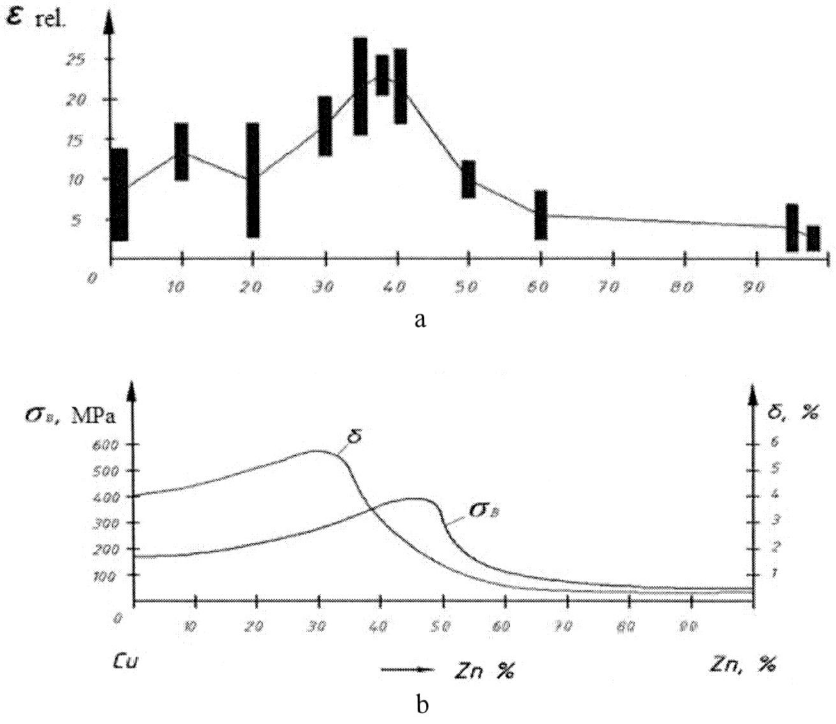

Figure 1.4. Dependence of properties of coatings from copper-zinc alloy on material composition: *a* – relative wear resistance ε of coatings; b – relative elongation δ and ultimate strength σ of material.

Depending on temperature and loading conditions, either a brass film 1 μm thick or a copper servovite film 0.5–2.0 μm thick may appear on the surface. It was found in [8] that the copper film contains micropores and transversal microcracks (figure 1.5) which serve as reservoirs for the lubricant.

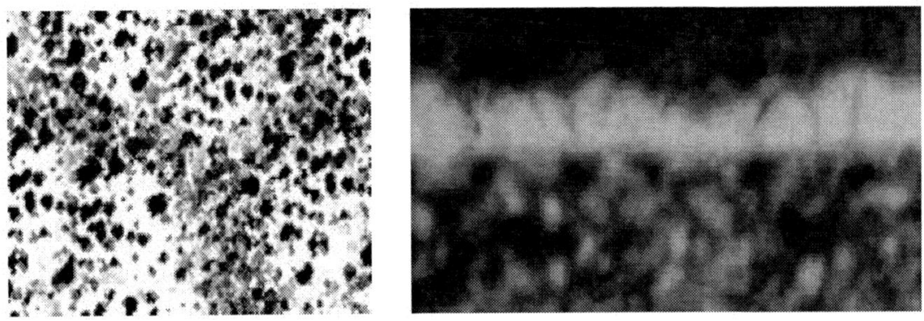

a *b*

Figure 1.5. Microstructure of servovite film (magnification ×1500): *a* – microporosity; *b* – transversal microcracks.

After the brass surface has been dissolved and both surfaces have become covered with a copper film up to 2 μm thick, glycerol stops dissolving atoms of alloying elements and the stationary stage of selective transfer begins.

In addition to the frictional rubbing of thin coatings, the electrofriction method is used in which the part is connected to the negative pole of a DC source and the rod is connected to the positive pole. The current density is 30–50 A/mm^2. The yield of the electrofriction coating deposition is 3–4 times higher than that of frictional rubbing. Electrofriction films have less porosity, higher adhesion to the base metal and longer life in friction compared to frictional films [6].

Frictional brass-plating was first applied in aircraft units to protect steel parts against seizure and scoring. The parts were coated that were fit with interference and dismantled periodically as well as the parts of heavy-duty but slow-moving joints. For plunger pairs of the fuel system of engines frictional brass-plating reduces the friction coefficient and increases resistance to seizure at axes misalignment. The brass-plating of hinge-bolt joints eliminates seizure occurring when steel contacts steel. Brass-plating is applied to prolong the life of hydraulic pump shafts, hydraulic cylinder rods, engine cylinder liners, crankshaft journals etc. [6].

Despite some progress in practical application, the wide use of finish antifriction abrasive-free treatment is hampered by the lack of scientifically grounded approach of the problem of the deposition of coatings and their effective usage. Selective transfer is applied in various domains of machine building, yet its relatively poor repeatability in many cases makes its practical use difficult.

1.2. Friction Surface Treatment Using Metal-Plating Lubricants

Under heavy contact loads, hence, at elevated temperatures in the friction zone the efficiency of lubricants decreases considerably that results in severe wear of friction units. One of the ways of widening the load and temperature ranges of operation of lubricants is to introduce additives containing powders of soft metals or their compounds into the lubricants.

The basic principle of action of such materials called metal-plating materials is the formation of protective films on real contact spots. The films localize shear deformations in friction and reduce the force interaction and wear rate of the contacting surfaces. Powders of copper, tin, lead, zinc, cadmium, and their alloys are most often used as soft metals [7].

Depending on the composition of lubricants, method of introducing additives into them and operating conditions various mechanisms of lubricity may act. Basic role in the decrease of the wear rate of friction members is mostly attributed to the formation of metallic films on a friction surface [9], yet in some papers high tribological properties are explained by the appearance of coordination compounds of metals added to the lubricant or formed in friction [10, 11]. This is also attributed to the synthesis of metal-polymer films in friction [10, 12].

By the phase compositions metal-plating lubricants are divided into heterogeneous and homogeneous. The former contain additives with a powdered metal or its oxides and the latter contain metal compounds soluble in the base oil. Heterogeneous lubricants are mainly used as the base for greases that is caused by poor sedimentation stability of metal powder dispersions. Fluid heterogeneous lubricants contain metal powders, copper most often, of nano-scale size (the particle size is 10–100 nm) and a stabilizing additive, e.g., oleic acid. Homogeneous lubricants possess unlimited stability but their practical application is restricted by some demands, first, to the rate of reduction of metal compounds in the friction zone which should provide the formation of a metal-plating film and, second, to the corrosion activity of the lubricants relatively to the friction pair materials.

Heterogeneous metal-plating lubricants are mainly used in pairs steel – copper alloy [9, 13]. At friction of steel or cast iron specimens in such lubricants the friction coefficients in the base oil and in the corresponding metal-plating material usually differ little. The introduction of a metal powder, e.g., copper can either reduce or increase the friction force [14]. Homogeneous metal-plating lubricants are mainly used in iron-based friction pairs such as steel–steel and cast iron – cast iron. It is explained by the necessity of the presence of an active reducing agent to form metal-plating films. Iron serves as the reducing agent for copper-containing additives used most often in such lubricants. At the same time, these lubricants provide a possibility of considerable decrease in the friction force and wear rate in other friction pairs, e.g., steel – copper alloy, by forming protective films based on copper contained in the copper alloy. In this case, like for heterogeneous lubricants, the introduction of the copper-containing additive may not only reduce the wear rate but also give some rise to the friction force [15].

Not any lubricant containing metal compounds can be considered as metal-plating. For example, the authors of [16-21] propose to introduce compounds of chrome, nickel, iron, and cobalt to oils in addition to salts or complex compounds of such soft metals as copper, tin, zinc, cadmium, and magnesium. Apparently, the films based on these metals are not responsible for decrease in the friction force and wear rate since their mechanical, hence, tribological properties differ little from those of the friction pair materials.

The comparison of the tribological behavior of heterogeneous and homogeneous metal-plating lubricants in steel–steel and cast iron – cast iron pairs with the results of study of heterogeneous lubricants in steel – copper alloy pairs shows the following regularity. Independently of the method of introducing the soft metal into the friction zone, hence, of the mechanism of the formation of metal-plating films, in friction of ferrous metals films protect the surface against wear, yet show poor antifriction properties. For example, at the sliding velocity of 0.8 m/s and under the pressure of 5 MPa the friction coefficient of the steel – cast iron pair lubricated with oil that contains copper chloride and diethyleneglycol is about 0.10–0.12 [22] which is typical for boundary lubrication. Probably, the efficiency of the antifriction action of metal-plating lubricants is much governed by the mechanical characteristics of the contacting surfaces because they are not totally separated by the metal-plating film. The friction coefficient can also be reduced significantly (from 0.10–0.12 to 0.015–0.040) in the steel–steel pair but this requires the surfaces to be totally separated by a film a few microns thick [23].

The limiting carrying capacity of metallic films appearing in friction of steel–steel pairs with metal-plating lubricants is also restricted (10–16 MPa). Data on their efficiency under heavier pressures are lacked and a high performance of the pairs is explained by the formation of polymer-like structures on the contacting surfaces rather than the effect of the metal-plating films. This occurs, for example, when a lubricant contains organic complexes of metals [10–12].

Antiwear and antiscoring properties of homogeneous metal-plating lubricants under severe friction conditions should be considered separately. Antiwear properties manifest themselves most fully under light pressures when metal-plating films may exist in contact for a long time without being damaged. Antiscoring properties manifest themselves under very severe conditions of friction when lubricants from other classes including organic are ineffective. Within the intermediate range of conditions of friction, i.e. when the pressure and contact temperature are high, yet not so much that boundary lubrication does not provide the efficiency of the unit the introduction of soluble compounds of metals, particularly, copper into the base oil deteriorates, as a rule, its antiwear characteristics.

A film of the plating metal may appear on the friction surface as a result of the reduction of plating metal ions by both iron contained in the contacting materials and lubricant anions. The reduction of metal ions by iron forms a basis for the mechanism of action of most metal-containing additives proposed for homogeneous metal-plating lubricants [10, 24]. The reduction of metal ions by the lubricant or by a proper anion, apparently, is more favorable to the tribological behavior of metal-plating lubricants. This is explained by lesser degree of chemical interaction between metal ions and the friction surface, hence, a milder corrosive effect of the lubricant on the surface. It follows, particularly, from paper [25] in which glycerol solutions of copper salts were studied. It was shown that the introduction of an active reducing agent for copper (bivalent tin salts) into a lubricant improves its antiwear properties and the carrying capacity of the film.

Irrespective of the nature of chemical reactions leading to the reduction of a metal in friction contact the formation of films from homogeneous lubricants is based on the processes of the crystallization of the metal from solutions of their salts or metal complexes. These processes begin from the appearance of a nucleus of the new phase or a crystallite followed by its growth. Therefore, metal-plating films are discrete by the nature of their formation and they can be considered as liophobic, i.e. having a great excess of the free surface energy, dispersed systems. The disperse phase is presented by particles of the reduced metal while the dispersed medium – by the lubricant or products of its transformations at friction. The tribological behavior of these films, the mechanism of their action and their limited carrying capacity can apparently be attributed to properties of dispersed systems [12].

The formation of a freely dispersed system from metallic particles in the friction contact predetermines weak energy interaction of the particles with the friction surface. This makes their long existence in contact, hence, the existence of the film, impossible. Thus, when using metal-plating lubricants in steel–steel and steel – cast iron pairs, keeping a necessary concentration of particles of the reduced metal in the friction zone in case a freely dispersed system has been formed requires a certain rate of the reduction of the metal from the lubricant [12]. In these pairs the contact reduction of metal salts on steel in the friction zone is a dominating reduction process which is accompanied by the oxidation of an equivalent amount of copper. Therefore, to provide conditions of the stability of a dispersed system the severe corrosive-mechanical wear of the steel friction surface should run during the reduction of metal salts from the lubricant. This is a drawback of homogeneous metal-plating additives.

Another possible mechanism of the lubricity of metal-plating lubricants in steel–steel friction pairs is the plasticization of the metal surface layer (the Rehbinder effect). The physical nature of this phenomenon is in the penetration of liquid surfactants into micropores and cracks on solid surfaces that facilitates their development owing to decrease in the work of the formation of new surfaces (decrease in the surface energy) [26]. A copper-containing film has a strong plasticizing effect on the steel surface like a film of any other metal. However, unlike fluid surfactants and corrosive media, it may induce the selective effect of plasticizing that occurs only on spots of real contact with steel, i.e. without the softening of the material subsurface layers. As the friction surfaces contact over profile asperities, prerequisites appear for the smoothening of the latter. Probably, decrease in the friction force in the steel–steel pair when adding metal powders into the lubricant is explained by exactly surface smoothening rather than the formation of a metallic film or a film of complex compounds on the surfaces. For example, papers [10, 11] report data on tenfold decrease in the microroughness of a steel part surface lubricated with grease TSIATIM-203 containing copper powder and a complexing agent (R_a = 0.075 μm) compared with the microroughness of the surface lubricated with the same grease without the additives (R_a = 0.7 μm). The friction coefficient in these lubricants was 0.074 and 0.098, respectively.

The application of additives containing metal powders in antifriction greases is restricted. The main drawback of greases with metal-plating components is a short life of their dispersed phase. To minimize this effect the plating components should be of a stable nano-size and have a special shell that protects them against oxidation.

Among recent studies on metal-plating compounds for lubricants we note the following. The authors of [27] solved the problem of forming a protective cermet film during the operation of a friction pair. The film has heterogeneous structure and possesses a high density, wear resistance, and adhesion to the friction surface; its friction coefficient after running-in is no more than 0.03–0.07. The plating concentrate contains a powdered metallic filler, surfactants, poorly soluble metal salts of dialkyl dithiophosphoric acid, and a base oil. Additionally, cyclohexanol is introduced in it as well as a mineral filler that is based on silicates. The silicate-based mineral filler contains natural minerals like layered hydrosilicates such as serpentinite and/or chlorite. Metal salts of dialkyl dithiophosphoric acid are presented by salts of zinc and/or tin, and/or molybdenum, and/or aluminum, and/or copper, and/or cadmium.

Metal-plating ultradispersed additives are also used in study [20]. The proposed metal-plating additive is introduced into oils, in particular, antifriction metal-plating lubricants based on oils applied in friction units of various machines and mechanisms. The additive contains the following components (wt.%): ultradispersed iron – 3–4; ultradispersed diamond – 0.7–1.0; industrial or motor oil – the rest. The engineering result of using the additive is the improvement of the wear resistance of friction pairs and decrease in their friction coefficient.

The metal-plating lubricant described in [28] is intended for using in friction units of various machines and mechanisms to improve their wear resistance. The metal-plating lubricant contains the ultradispersed copper powder and oil. Its use reduces the friction coefficient twice and increases the rate of plating 3.5 times.

The grease containing highly dispersed additives [29] can be consider as a metal-plating grease used in friction units of wheel and caterpillar vehicles, machinery and marine mechanisms. The lubricant contains grease Litol-24 as a base and 5–20% of highly dispersed metallic powder. Powders of highly dispersed zinc, bronze, or lead are used. The metallic powder may contain alloying additives of stibium, tin, or cadmium. The grease possesses a good lubricity which provides the wearless operation of friction pairs and recovers worn friction surfaces while retaining the main technological characteristics of the grease.

Patent [21] describes the lubricant used to improve the fatigue strength and wear resistance of heavy-duty friction units, e.g. rolling bearings. It contains the powdered filler including nano-sized powders of iron, nickel, and zinc with the granularity of 10–30 nm. The composition of the lubricant is as follows (wt.%): the filler – 0.5–1.5; the grease – 98.5–99.5. The lubricant demonstrates metal-plating effect; its application improves the fatigue strength and the wear resistance of friction units.

The lubricant described in [30] contains soap grease and the powders of zinc and molybdenum disulphide. The addition of 4-13 wt.% of the zinc powder improves considerably, in authors' opinion, antiwear and antiscoring properties of the grease while the addition of 1–5 wt.% of molybdenum disulfide favors the most effective increase in its antifriction properties. However, when discussing the joint application of zinc and molybdenum disulfide in lubricants, we note that these components affect quite oppositely their tribological properties. Under heavy loads zinc demonstrates metal-plating effect that requires juvenile surfaces free of oxide films, surfactants etc. to exist. At the same time, molybdenum disulphide passivates the surfaces and reduces their surface energy thus hampering direct contact between active zinc and the friction surface.

Patent [31] describes the metal-plating lubricant that contains such known components as the powder of the copper–tin alloy with the particle size of 0.01–5.00 μm (4–12 wt.%), monocarboxilic fatty acid C_{12}–C_{18} (0.19–0.50 wt.%), hydrocarbon fuel (15–40 wt.%), and mineral oil (the rest). Additionally, it includes the activator that improves the adhesion of the surfaces of the metallic friction pairs to fine metallic particles contained in the lubricant. Inorganic and organic phosphates, e.g., disodium phosphate Na_2HPO_4 and tributyl phosphate $(BuO)_3PO$ may serve as an adhesion activator. Phosphates form a protective film on the surfaces of metal defects that prevents this surface against oxidation for a long time that provides a stronger adhesion of the metal powder to surface defects. The group of surface adhesion activators includes also nitrogen-containing heterocyclic compounds such as triazobenzene or ocridine bases that form stable film complexes with ions of the metal being dissolved.

The analysis of physical-chemical and tribological properties of metal-plating lubricants governing their performance shows that the lubricants are unsuitable for friction units which experience oxidative wear, i.e. for most ball and roller bearings. Their application may be advisable for heavy-duty friction units, especially sliding units, having a limited life and undergoing mainly seizure as well as for units in which the lubricant plays the role of a carrier of such additives. The use of additives of metal powders to antifriction greases is also restricted by probable shortening of the durability of the grease dispersed phase owing to its abrasion by plating components [32].

1.3. Treatment of Friction Surfaces by Recovering Compositions Based on Natural Layered Silicates

Various minerals and other compounds of geological origin become more widely spread recently as antiwear additives to lubricants [33–37]. Research in this field are aimed at the development of special additives to fuels and lubricants; they should be based on cermet compounds and should be capable of reacting with friction surfaces to form a cermet layer on them that partially heals surface defects and possesses high antifriction and antiwear properties. Such materials based on ground and modified serpentinite and other natural and artificial minerals are called geomodifiers and the compositions on their base are called recovering compositions or revitalizers. Many cermet materials are mixtures of the classic magnesia-ferrous silicate (serpentinite $Mg_6(Si_4O_{10})(OH)_8$) which is a form of a class of mineral ores belonging to olivines. The end phases of this mineral are forsterite (Mg_2SiO_4) and fayalite (Fe_2SiO_4) and also silica (SiO_2) and dolomite ($CaMg(CO_3)_2$) in small amounts as well as attendant inclusions of Fe, Ca, Ni, Ti, Cr, Cu, etc. in the form of oxides and other compounds.

To obtain the required effect of using the geomodifier it should destruct according to the following reaction:

$$Mg_6\{Si_4O_{10}\}(OH)_8 \rightarrow 3Mg_2\{SiO_4\} + SiO_2 + 4H_2O \text{ (at } t = 600\ ^\circ\text{C).}$$

An organic cermet film appears on the contacting surfaces during operation of a friction pair. It is a liquid monocrystal grown on the lattice of the surface layer of the metal itself. At the same time, the effects of embedding, diffusion, and segregation provide the structure transformation of the subsurface layer owing to the migration of atomic hydrogen towards the zones of elevated temperatures (the subsurface layer has the maximal temperature because of plastic deformations resulted from contact between the surfaces).

The processes occurring during lubrication with a lubricant containing geomodifiers are similar to those known in the metallurgy of ferrosilicates (Fe-Si and other siliceous alloys like Ca-Si, Ca-Si-Al etc). Under the effect of the mentioned factors reactions of oxidation and reduction run on a steel surface. To create stable equilibrium in a tribosystem required for the reactions to run it is necessary to introduce into the friction zone hydroradaicals containing catalysts, i.e. ions of metals with variable valence. Such conditions hamper the formation of free radicals and owing to this metal ions are kept in the friction zone inhibiting the wear of the contacting surfaces. Water is a by-product of the reactions of oxidation and reduction when their rates vary. The hydrophase formed is then involved in the reduction of the friction force and wear rate of a tribosystem.

The chemical and phase compositions of the metal hydrosilicates used are due to the presence of complex conglomerates of octahedral and tetrahedral compounds with the bonds Si–O–Si, Si–O–OH–M etc. During mechanical and thermal loading some bonds are ruptured and compounds are formed having the free bonds like H–O–(...), H–O–OH–(...) as well as water. This is accompanied by the active replacement of bonds owing to the adsorption of hydrogen and new bonds like Si–O–OH, Si–O–Fe etc. appear. This process induces the following reactions between crystals of the silicate-ceramic composition and crystals of the metal phases:

$$Mg_6(Si_4O_{10})(OH)_8 + Fe_2O_3 + H_2 \rightarrow 4(MgFe)SiO_4 + 5H_2O.$$

The protective cermet layer results from the reactions of the substitution of magnesium ions in crystals contained in geomodifier minerals by iron ions from the surface layer of the friction surfaces.

Figure 1.6 shows the principle of the technology of the recovering worn surfaces using nano-ceramics.

The tribological assessment of geoactivators is performed with account for their capability of inducing micrometallurgic processes which yield surfaces of metal silicates similar to forsterites or olivines. This means that initially a geomodifier should correspond to certain values of energy density and activity to hydrogen and water.

Before the moment of decomposition, serpentine acts as a common abrasive. After the decomposition of the geomodifier ceramic and cermet particles contained in it enter the cleaned friction zone together with a catalyst. The contact zone becomes depleted in free hydrogen; the structure of surface layers changes and their strength increases several times due to diffusion, the Kirkendall effect and segregation. During subsequent operation the friction surfaces become covered with an organic cermet coating that heals partially defects on the friction surface and has high antifriction and antiwear properties.

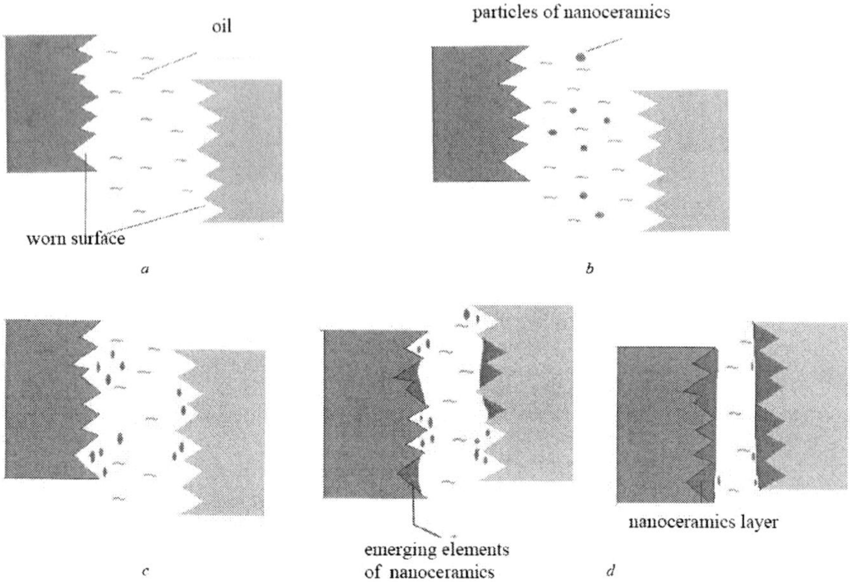

Figure 1.6. Principle of functioning of geomodifier: *a* – original state of friction surface; *b* – introduction of nano-sized ceramic elements in friction zone; *c* – agglomeration (concentration) of ceramic nanoparticles in worn areas of friction surface; *d* – formation of nanoceramic layer.

The authors of [38] used a mixture of artificial hydrosilicates of nickel, i.e. diatomite and pearlite, as a geomodifier of friction. They have layered structure. Diatomite and pearlite belong to subclass VI of dimeta silicates with layered radicals. In this case the radical is an infinite layer consisting of sextuple rings of silica-oxygen tetrahedrons whose free apices are directed similarly; Si^{4+} is often (up to half of the ions) replaced by Al^{3+}. They possess perfect cleavage, ductility, and sharply decreased hardness, i.e. they are wax-like. An additional ion $[OH]^-$ or F^- also necessarily presents (one ion per a ring) as well as layers of additional ions $[OH]^-$. For example, they are contained in serpentine $Mg_6\{Si_4O_{10}\}(OH)_8$ whose formula can be presented as follows: $Mg_3[OH]_2\,Si_4O_{10}\cdot 3Mg[OH]_2$.

Paper [39] describes the antifriction and antiwear suspension including the dispersion medium, i.e. a monomer with unsaturated bonds, and a metal silicate. The use of monomers with unsaturated bonds as the dispersion medium allows a protective antifriction tribopolymer film to appear on the surface of any metal almost immediately. It becomes protecting the contacting surfaces right after the introduction of the suspension while the above-described compositions require a run-in period to pass after which they start demonstrating antifriction effect.

Both one and several metal silicates can be introduced into the dispersion medium as fine additives (the grain size is 1 to 10 microns). The selection of a certain metal silicate or a combination of metal silicates depends on the contacting materials, their wear or on the composition of a modified layer that is required to appear on the friction surfaces. The components introduced into the dispersion medium are most often natural silicates of Mg, Al, Zn, Ni, Li, and Ba. Increase in the content of a metal silicate in the dispersion medium reduces the time required for surface modification to occur and prolongs the period during which the modified surface shows antifriction and antiwear properties.

The authors of [39] distinguish the following phases of the influence of the suspension on part surfaces:

- the formation of a protective film, the cleaning and microgrinding of friction surfaces as well as the embedding of a metal silicate into them owing to the contact pressure produced by the mated parts, e.g., two gear teeth, the ball and race in the rolling bearing, a compression ring and liner wall etc.;
- the distribution of the metal silicate over a subsurface volume, hence, the recover of the shape and dimensions of the parts;
- the provision of the antifriction and antiwear effect on the contacting surfaces owing to decrease in their roughness to $R_a = 0.15$ μm and the equating of the hardnesses of both surfaces to HRC 56–60;
- the creation of conditions for the appearance of a tribopolymer film on the part surfaces and improving its adhesion.

Paper [40] describes the composition used to form a new layer on the rubbing metallic surfaces which contains a fine base consisting of natural nickel-iron-magnesia hydrosilicates with the generalized chemical formula $\{(Ni_3);(Fe_3);(Mg)\}(Si_2O_5)(OH)_4$ and a catalyst, i.e. the mineral belonging to the group of olivines – forsterite (Mg_2SiO_4) or fayalite (Fe_2SiO_4). The grain size of the base and catalyst is 1 to 100 μm. The proposed composition provides the formation of a new layer capable of failing during the friction of the contacting surfaces and then restoring. In the original base, the components involved subsequently in the formation of a new layer already make up the united crystalline structure (lattice). Therefore, the initial preparation of the base includes only its dispersion down to a grain size suitable for introducing it into the contact zone. The presence of the catalyst allows the process of mass exchange between the contacting surfaces to start. The use of the proposed composition leads to some wear of the new surfaces formed under its effect. This is the condition necessary and sufficient for the simultaneous running of self-oscillating "pendulum" mass exchange processes between such new friction surfaces belonging to one and the same closed tribosystem. This is achieved owing to nano-sized particles (down to 0.1 μm in size) appearing during the wear of the newly formed layer. The particles are apt to ionization and polarization relatively to the parent surface base. This newly acquired property of the particles orients them in the closed space of the existing tribosystem in such a way that they approach to the surface being opposite to the parent base. In this case the lubricant that presents between the friction surfaces plays the role of a third body carrying the particles. It also provides the closeness of the system hampering the effect of the environment on the particles liberated in friction. Provided that physical-chemical properties of the material of the newly formed layer and the particles separated from it are relatively uniform and their structural compositions are identical and that the activity of the particles is high due to their gross surface energy and heavy pressures in microvolumes in which the particles come during the contact of the surfaces of the closed tribosystem, the particles immediately become involved in the reaction of joint crystal formation with the newly formed surface opposite to the parent one. Correspondingly, similar particles separated from the counterface tend to replace them. The process stops when friction ceases and begins again with a new friction cycle.

Thus, the recover and hardening of movable joints by cermet materials are performed owing to the formation of structures with an increased strength on friction surfaces, the inhibition of the processes of hydrogen wear and embrittlement of a metal and the improvement of the thermodynamic stability of the system "friction surface – lubricant". High mechanical, tribological, and physical characteristics of the surface (the microhardness is up to 690–710 HV, the toughness is up to 50 kg/mm^2, the friction coefficient is 0.003–0.007, and the heat resistance is up to 2500 °C) allow the developers of geomodifiers of friction to recommend them for the recover of friction units of various machines and mechanisms, e.g., combustion engines, equal angular velocity joints, reducers and transmissions, oil pumps and hydraulic motors, compressors, machine-tools, pneumatic hammers and presses, smoke suckers and ventilators, open gearings and chain-drives etc.). The life of friction units is prolonged 2–3 times, energy and fuel consumption decrease by 10–15% and maintenance cost also reduces. The recover and hardening of friction surfaces are performed during the operation of a friction unit and do not require it to be disassembled.

However, we note that a number of serious problems arise when using recovering compositions based on natural silicates. In particular, a positive effect of the application of these compositions occurs mainly for heavy-duty steel surfaces. When using the geomodifier in a “pair soft material – hard material” the pair wear increases due to the embedding of non-decomposed coarse particles of the geomodifier into the softer material and their abrasive effect on the rubbing surfaces.

When cermet materials are used for surface treatment, a great amount of free water is liberated which decomposes thus intensifying the hydrogen wear of other parts of a machine or mechanism.

If a unit is treated with a geomodifier, its temperature regime is distorted because of heat removal from the friction zone resulted from an additional thermal resistance in the form of a cermet layer with a low thermal conductivity.

Another problem relating to the mass production and wide application of recovering compositions based on natural layered silicates concerns the unstable quality of natural raw, hence, poor predictability of results of the use of geomodifiers.

The above leads us to a conclusion that geomodifiers and technologies of recovering friction units based on them are highly effective. However, many unsolved problems exist that hamper their wide application.

REFERENCES

[1] Kershenbaum, V.Ya. Mechano-Thermal Formation of Friction Surfaces; Mashinostroenie: Moscow, SU, 1987; 232 p.

[2] Improvement of Wear Resistance on Basis of Selective Transfer; ed. Garkunov, D.N.; Mashinostroenie: Moscow, SU, 1977; 215 p.

[3] Garkunov, D.N. Triboengineering. Wear and Wearless Friction; Moscow Agricultural Academy: Moscow, RU, 2001; 616 p.

[4] Polzer, G., Firkovsky, A. et al. Finish Antifriction Abrasive-Free Treatment (FAAFT) and Selective Transfer; Durability of Friction Members of Machines; Mashinostroenie: Moscow, 1990; Vol 5, pp 85-122.

[5] Polzer, G., Meißner, F. Grundlagen zu Reibung und Verschleiß; VEB Deutschter Verlag für Grundstoffindustrie: Leipzig, DE, 1983; 264 p.

[6] Andreeva, A.G., Burumkulov, F.Kh. et al. Finish Antifriction Abrasive-Free Treatment as Way of Prolonging Life of Machinery; Durability of Friction Members of Machines; Mashinostroenie: Moscow, SU, 1990; Vol 4, pp 34–59.

[7] Friction, Wear, and Lubrication (Tribology and Triboengineering); ed. Chichinadze, A.V.; Mashinostroenie: Moscow, RU, 2003; 576 p.

[8] Lukashok, A.N., Tikhonov, P.V. Two-Layered Nanocrystalline Coating for Precision Sliding Pairs; Advanced Materials; *Interkontakt Nauka:* Moscow, RU, 2007; pp 294 -298.

[9] Selective Transfer in Heavy-Duty Friction Units; ed. Garkunov, D.N.; Mashinostroenie: Moscow, SU, 1982; 207 p.

[10] Kuzharov, A.S., Bulgarevich S.B. et al. *Journal of Friction and Wear.* 2001, Vol 22, 650-658.

[11] Polyakov, A.A. Effect of Wearless Friction and Tribotechnologies. 1996, Vol 3-4, 47-122.

[12] Komarov, S.N., Pichugin, V.F. et al. Metal-Plating Lubricants for Steel–Steel Friction Pairs; Durability of Friction Members of Machines; Mashinostroenie: Moscow, SU, 1990; Vol 5, pp 70-85.

[13] Patent 958479 (SU)

[14] Melnichenko, I.M., Gribailo, A.P. et al. *Journal of Friction and Wear.* 1980, Vol 1, 674-677.

[15] Patent 859425 (SU)

[16] Patent 3256188 (US)

[17] Patent 825592 (SU)

[18] Patent 808527 (SU)

[19] Patent 1154316 (SU)

[20] Patent 2178803 (RU)

[21] Patent 2258080 (RU)

[22] Patent 1086009 (SU)

[23] Belov, P.S., Komarov, S.N. et al. *Journal of Friction and Wear.* 1984, Vol 5, 1033-1039.

[24] Patent 595336 (SU)

[25] Patent 1171511 (SU)

[26] Rehbinder, P.A. Surface Phenomena in Dispersed Systems. Physical-Chemical Mechanics. Selected Papers; Nauka: Moscow, 1979; 381 p.

[27] Patent WO2005071049 (RU)

[28] Patent 2054030 (RU)

[29] Patent 2139920 (RU)

[30] Patent 2161177 (RU)

[31] Patent 2139319 (RU)
[32] Ischuk, Yu.L. Composition, Structure, and Properties of Greases; Naukova Dumka: Kiev, UA, 1997; 512 p.
[33] Kobylskov, I.P. Complex Solutions of Problems of Energy Saving and Reliability in Operation of Machines Based on Recovering Technology; Innovations in Machine Building; Joint Institute of Machine Building of Belarus National Academy of Sciences: Minsk, BY, 2008; pp 125-128.
[34] Balabanov, V.I., Beklemyshev, V.I. et al. Friction, Wear, Lubrication, and Self-Organization in Machines; Izumrud: Moscow, RU, 2004; 192 p.
[35] Patent 2169172 (RU)
[36] Patent 2168662 (RU)
[37] Patent WO2007/082299 (US)
[38] Patent 10897 (BY)
[39] Patent 2237704 (RU)
[40] Patent 22266979 (RU)

Chapter 2

LUBRICANTS MODIFIED BY NANO-SIZED HARD COMPONENTS

2.1. BASIC PRINCIPLES OF DEVELOPING GREASES MODIFIED BY NANO-SIZED COMPONENTS

The life of friction units of machinery is largely governed by the performance of lubricants used. Progress in engineering is concerned with increase in velocities and loads in friction units that requires new lubricants having an improved carrying capacity and providing a lower wear rate. In heavy-duty friction units, especially in sliding joints, grease is almost completely squeezed out of the friction zone irrespective of the nature of its thickener and base oil. In such cases hard additives play an important role, especially nano-sized ones.

At present the following groups of additives to greases are known to realize elements of nanotechnologies.

Antifriction and Run-In Compositions Based on Nano-Diamonds

Nano-diamonds with the grain size of 4–6 nm and cluster carbon contained in them structure an oil film and increase its dynamic strength as well as affect the lattice of the surface metal thus hardening it, form new friction surfaces and reduce boundary friction and wear, especially under heavy loads. As a result, running-in shortens and the quality of friction units becomes optimal [1–12].

Nano-Conditioners of a Metal

As a result of tribochemical reactions of the formation, decomposition, and reduction of compounds of a metal with active molecules of the product in the friction zone, such conditioners form a protective boundary layer 20–40 nm thick. The protective layer acquires plastic and elastic properties as well as antifriction properties and resistance to heavy loads [13–16].

Reconditioners

Reconditioners form the above-mentioned protective layers and favor increase in the carrying capacity (strength) of an oil film. The polymolecular system of such compositions includes nano-sized complexes or clusters of organic compounds. It structures the boundary oil film and improves the adhesion of a lubricant to the metal [17–22].

Recovering Additives and Remetallizers

These are oil-soluble or powdered metal-organic compounds. They realize the tribochemical ("ion") mechanism of the metal-plating of friction surfaces due to the formation or recover of a metal-containing nano-structured protective film on them. The additives favor the healing of microdefects on friction surfaces and recover their performance [23–28].

Geomodifiers (Geoactivators)

These are additives based on natural or artificial minerals with nano- or microparticles. Geomodifiers are deposited on friction surfaces as they are carried by oil; they induce the formation of a cermet coating with a high wear resistance and a low friction coefficient [29–33].

The improvement of tribological properties of a friction unit lubricated with a lubricant including nano-sized diamond-containing components results from changes in physical-chemical and rheological properties of the lubricant (in particular, a higher thermal stability and carrying capacity of an oil film [34, 35]) as well as from the hardening of surface layers of the friction members due to their severe plastic deformation during lubrication by the lubricant containing hard particles.

One of possible mechanisms of the antiwear effect of nanomodifiers is based on the ordering of the lubricant structure under the action of the self-charge of nanoparticles. Dispersed particles of various origin acquire a charge because of their imperfect structure resulted from external factors which manifest itself at the stage of particle production or from factors arising during contact as a result of triboactivation and tribofailure. The self-charge of modifier particles induces electrophysical processes in the lubricant that affect its structure. Under the effect of the force field of nanoparticles an oriented layer is formed from the dielectric component of the lubricant [36]. The layer thickness depends on the charge of nanoparticles and peculiarities of the structure of lubricant molecules; it can reach 0.02–0.04 μm [37].

The similarity of the mechanisms of the effect of nanoparticles of various origin on the structure of liquid-phase media manifests itself as the similarity of the stabilizing effect of carbon nanoparticles, nanopowders of oxides, metals and other components on the rheological behavior of lubricants [38].

If a lubricant contains active antiwear additives, nanomodifiers are capable of providing the synergic effect of the transfer and adsorption of molecules of the friction modifier [39].

Under optimal regimes of the operation of a tribosystem it demonstrates self-organization [40, 41]; in this case the duration of steady-state wear is quite long. The self-organization of the tribosystem manifests itself in the formation of an adsorbed layer of the lubricant of juvenile surfaces; the layer possesses a high temperature and deformation resistance compared to the similar layer appeared on the original friction surface. Thus, owing to the joint impact of various physical-chemical phenomena the tribosystem forms a favorable cycle that retards the wear process. Here one should take into account that cycles of the physical-chemical processes are also formed as a result of such negative contact phenomena as temperature flashes on real contact spots, the effect of abrasive particles, fatigue wear, and corrosion damage.

Greases are colloid systems consisting of a dispersion medium (mineral or synthetic oil) and a disperse phase (salts of high-molecular acids). They are widely applied to lubricate nonhermetic friction units providing the simplification of the unit design and favoring decrease in the lubricant consumption. However, under heavy loads the lubricant film breaks and metallic contact between the friction surfaces occurs that is accompanied by the squeezing of the grease out of the friction zone. This causes the deterioration of antifriction characteristics of the pair, increase in the wear of the friction surfaces, disturbance of the operating conditions of the pair and may lead to fail due to seizure and scoring.

Greases consist of three main components, i.e. a disperse phase (5–30 wt.%), dopes (1–5 wt.%) and a dispersion medium (the rest). The dispersion medium is natural or synthetic oils.

The disperse phase is formed by a hard thickener and appears when salts of high-molecular fatty acids are introduced into oils.

Dopes are additives, fillers, and structure modifiers. Additives are divided into antioxidants and antifriction additives. Fillers and modifiers are hard disperse compounds that are almost insoluble in the dispersion medium and form a separate base.

Thus, greases hold an intermediate position between oils and solid lubricants. They belong to structured colloid systems. Their properties depend on peculiarities of the three-dimensional structure frame formed by the disperse phase. Cells of the frame hold a great amount (80–90%) of the dispersion medium. The stability of the structured system depends on the strength of the structure frame, forces of interaction of its single particles and between its elements and the dispersion medium at the interphase, the critical concentrations of various soaps and other colloidal-chemical factors.

Recently more and more attempts are made to improve directionally properties of greases by introducing the third component into them, i.e. a dope. The use of dopes to greases, primarily, the joint introduction of additives and fillers allow developers to vary the grease structure most pliably. Additives affect strongly the formation of the grease structure and play a slight role in the construction of its single elements. Fillers are also elements of the grease structure though their effect on its formation is weaker compared to that of additives. The principal difference between fillers and additives is the insolubility of the former in the dispersion medium. At the same time, the community of purposes and colloidal-chemical principles of controlling the structure and properties of greases due to additives allows one to consider their use as a united trend of improving grease performance.

Various hard additives contained in a grease are retained in the friction zone even if the grease is squeezed out of it. They form a separating layer preventing the metallic surfaces against direct contact. The structure and properties of such layers depend on the nature of hard additives and the mechanism of their effect.

Works related to the use of various nano-sized components as hard additives have recently become of still greater scientific and practical importance. Their application yields principally new results compared to those of using previous additives.

Nano-sized additives find application in both oils and greases. However, in the latter case problems arise related to the colloid structure of greases. This specific feature of the structure of the disperse phase of greases predetermines their rheological behavior and hampers much the homogeneous distribution of additives in the grease bulk. This is especially important for the nano-sized additives whose grain size is comparable with the size of fibers that form the structure frame of soap greases. It is difficult to achieve the homogeneous distribution of nano-sized additives over the grease bulk without decomposing the structure of its disperse phase.

Nanoparticles producing the modified effect on lubricants are presented by particles of metals, ceramics, natural silicates, graphite, fullerenes and fullerene-like structures, and carbon-containing products of detonation synthesis such as ultradispersed diamonds (UDD) and ultradispersed diamond-graphite charge (UDDG).

Nanoparticles introduced into a lubricant produce a complex effect on a tribosystem in which both favorable and undesirable physical-chemical processes run. Among the processes that prolong the tribosystem life are the formation of separating layers with an improved carrying capacity, the provision of the stability of rheological characteristics of a lubricating film within a widened temperature range, decrease in the friction surface roughness and the hardening of surfaces layers [42–48].

However, in some cases antiwear and antifriction properties of lubricants deteriorate owing to their tribodestruction and oxidation. Aggregated structures are formed from wear particles that intensify abrasion, the severe wear of friction surfaces by clusters of nanoparticles occurs, the structuring of a lubricant runs that increases its viscosity and the corrosion effect on surface layers of the friction members intensifies [49–51]. Such multifunctional effect of modifying nanoparticles requires conditions for the primary realization of favorable processes in the tribosystem which would improve antifriction properties of friction pairs and reduce their wear rate.

With account for possible mechanisms of the operation of a tribosystem containing nanoparticles [52] we can formulate the following principles of developing lubricants modified by nano-sized components:

- the introduction of adsorption-active nanomodifiers into lubricants in order to form a separating layer on a friction surface that is characterized by an improved carrying capacity and a widened range of working temperatures;

- the application of nanomodifiers to transport functional additives into the friction zone;
- the introduction of nanomodifiers into lubricants to provide the running of microcutting, the removal of a defect layer from a friction surface and the appearance of juvenile surfaces;
- the use of hard nano-sized components favoring the running of the severe plastic deformation of surface layers of the friction members followed by the formation of nano-sized cell structures that have a higher resistance to the nucleation and propagation of microcracks and absorb effectively the frictional energy.

2.2. Developing Method of Formation of Structure Frame of Grease Having Improved Characteristics

The use of ultradispersed diamonds as a modifier for lubricants is explained by the following factors: nanodiamonds structure an oil film, increase its dynamic strength, affect the lattice of the metal surface and harden it, form new friction surfaces and reduce boundary friction and wear, especially under heavy loads.

These properties of ultradispersed diamonds are realized by many developers of lubricants to improve their antiwear and antifriction properties and widen the ranges of working temperatures and loads.

Paper [9] describes the composition of a grease containing the following components (wt.%): the ultradispersed diamond-graphite powder – 0.2–0.5, the highly dispersed metal salt – 2–15, and the soap grease – the rest. The highly dispersed salt can be presented by stannous sulfate, copper sulfate, barium sulfate, lead sulfate, and barium sulfide. The grease is intended to lubricate friction surfaces of main units of machines and mechanisms. Its use provides the improvement of the performance of mechanisms and prolongs the life of machines due to its high antiwear and antifriction properties.

The authors of [8] applied the ultradispersed diamond-graphite powder as an antifriction run-in additive. The additive is intended for automotive, tractor, marine, and airplane motors; it also can be used to prolong the life of machinery in manufacturing industry. The additive composition is as follows (wt.%): 0.1–0.5 of the hard modifier, i.e. the ultradispersed diamond-graphite charge and mineral oil (the rest). The ultradispersed diamond-graphite charge has the primarily grain size of 4–6 nm and the specific surface of 400–500 m^2/g. The content of diamond in the charge is no less than 30 vol.%. As a result, the duration of the running-in of parts shortens, the sedimentation stability of the antifriction additive increases and the toxicity of exhaust reduces.

Paper [7] describes the lubricant containing base oil with the additive being a natural diamond micropowder (1–10 wt.%). The effect of its application includes the improvement of the lubricant performance, viz. its antiscoring properties. The commercial natural diamond powder was added to this lubricant that required thorough inspection of its homogeneity including the control of the particle size and composition. The antiwear and antiscoring properties of the lubricant increase while the rest performance characteristics, i.e. the mechanical, chemical, and colloidal stability, penetration, viscosity, pumpability, deteriorate. It should also be noted that, as the lubricant contains 1–10 wt.% of hard additives, it can be considered as a lubricating paste rather than a grease.

The authors of [3] also propose to use additives of the nanodiamond powder as a run-in component of lubricants. The developed grease contains 10–15 wt.% of sulfur, 10–20 wt.% of molybdenum disulfide, 5–7 wt.% of indene-coumarone resin, 1–15 wt.% of the ultradispersed diamond powder or ultradispersed diamond-containing charge, and stearic acid (the rest). The grease provides the efficient run-in abrasive treatment of metallic friction surfaces.

In addition to good run-in properties, ultradispersed diamond particles produce a strong structuring effect due to their high surface energy and the compatibility of carbon clysters with base oil. This property is used when developing greases with a branchy structure frame that possesses a high oil-holding capacity and increases the colloidal stability of the grease and the carrying capacity of a lubricating film.

In the above-mentioned study the problem of improving antiscoring properties and the colloidal stability of the complex lithium grease is solved by the additional introduction of 0.4–0.5 wt.% of ultradispersed diamonds into it. The introduction of the ultradispersed diamond-graphite charge into the dispersion medium (oil) and its disintegration are performed before acids are added in it at the oil temperature of 70–80 °C. The disintegration lasts until particles become 10–30 nm in size [2]. The diamond-graphite charge is produced by detonation synthesis and contains 60–70% of diamonds covered with a graphite-like shell. It should be noted that UDDG particles, like all nanoparticles, have a great specific surface (400–500 m^2/g) and a high surface energy that leads to their agglomeration under normal conditions. Original UDDG particles 4–6 nm in size form fractal cluster structures 30–40 nm in size that make up a larger (0.4–0.5 μm in size) easily decomposed aggregates [52–55]. Therefore, to recover the high activity and structuring capacity of UDDG particles it is necessary to disintegrate agglomerates before acids are added to the dispersion medium.

Let us consider the formation of the structure of the grease disperse phase in the presence of a nano-sized modifier using lithium grease Litol-24 with ultradispersed diamond-graphite charge as an example.

The technology of lithium greases with nano-sized additives is developed by the authors of [35, 56] and includes the basic stages that are presented in figure 2.1.

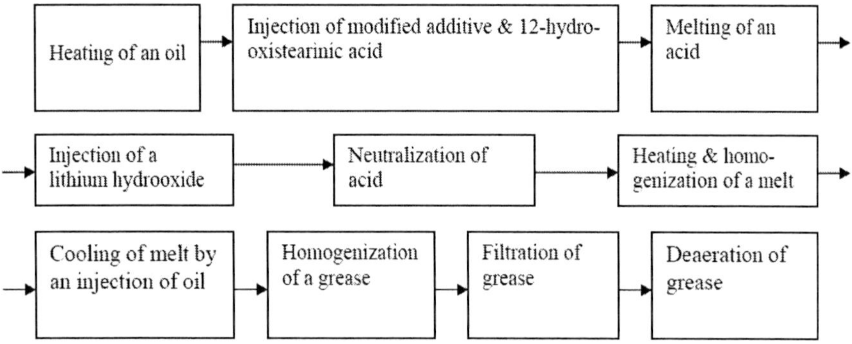

Figure 2.1. Stages of fabrication of modified lithium grease.

The structure frame of the greases was examined using a scanning electron microscope Cam Scan (Oxford Co., GB) and a high-resolution scanning electron microscope Mira (Tescan, Czechia). The dispersed phase was flushed following the procedure described in [57].

Introduced UDDG particles play a role of centers of structuring of the grease dispersed phase. The process of complexation begins to run on surfaces of UDDG particles that increases the degree of structuring of the dispersed phase and the thickening ability of complex lithium salts as well as favors the formation of a branchy structure frame (figure 2.2). This improves the oil-holding capacity of the frame and increases the strength of a boundary lubricating film in the friction zone. UDDG particles are implanted into the dispersed phase rather than remain in the grease dispersion medium. This, first, increases the mechanical and thermal stability of the grease and, second, prevents UDDG particles against agglomeration thus improving the grease colloidal stability. The structure frame of the grease consists of shorter fibers 5–10 μm long with a great number of interweavings; this increases its colloidal stability.

The influence of the content of UDDG particles on properties of the grease was studied when estimating the carrying capacity and colloidal stability of the modified complex lithium grease. The seizure load and welding load were determined using a four-ball friction machine according to GOST 9490. The colloidal stability of the grease was determined according to GOST 7142. Six compositions of the grease were tested with different content of UDDG and the same content of other components. The test results are presented in Table 2.1.

Table 2.1. Antiscoring Properties and Colloidal Stability of Greases

Number of composition	Content of UDDG, wt.%	Seizure load, N	Welding load, N	Colloidal stability, %
1	-	784	1568	12
2	0.3	784	1744	7
3	0.4	872	2195	5
4	0.5	872	2450	5
5	0.7	696	2450	2
6	1.0	617	2195	1

It has been found that the grease with the optimal content of the UDDG additive has better characteristics than the non-modified grease. In addition to increase in the seizure load, welding load and colloidal stability of the grease, the UDDG additive raises the hardness of the contacting surfaces and favors decrease in the friction coefficient owing to surface smoothing.

Figures 2.3 and 2.4 show time variations in the friction coefficient of the pairs annealed steel 45 – hardened steel 45 and hardened steel 45 – hardened steel 45 lubricated with the grease Litol-24+UDDG prepared according the following procedures:

- technology 1: the introduction of modifying particles into the prepared grease (figure 2.3);
- technology 2: the implantation of modifying particles into the structure of the dispersed phase (figure 2.4).

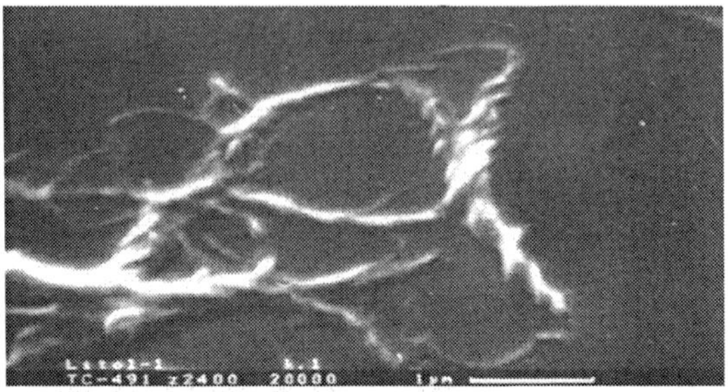

a

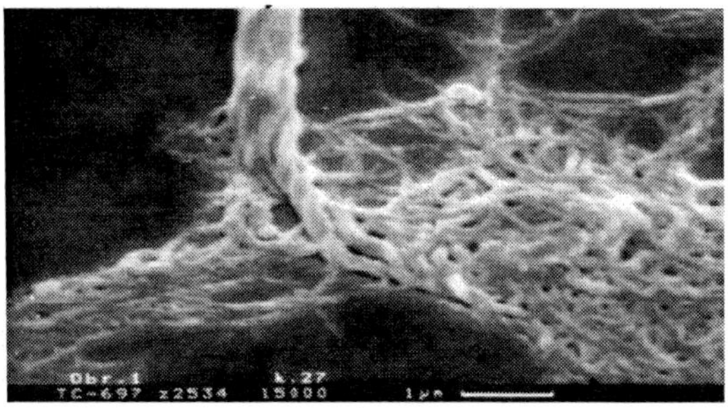

b

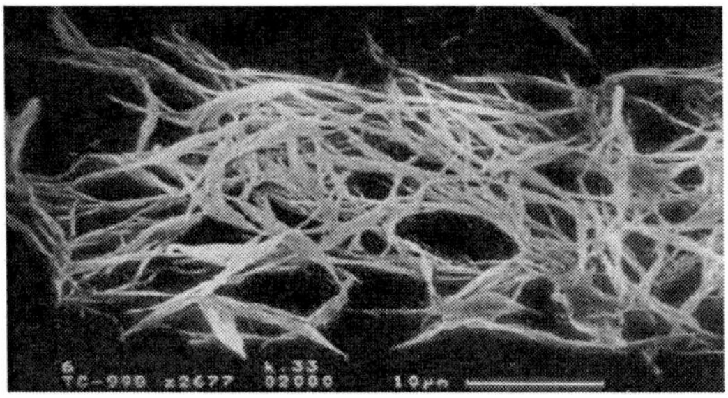

c

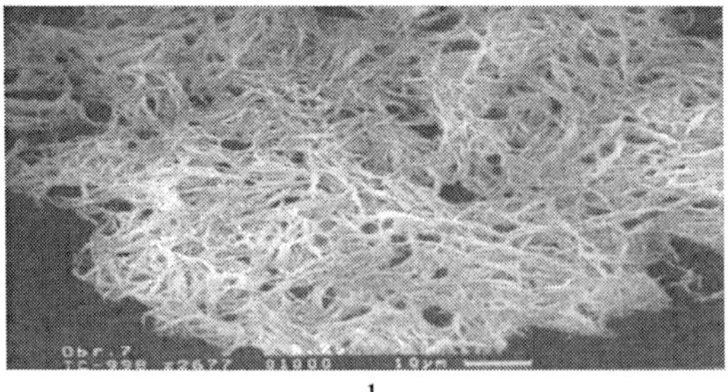

d

Figure 2.2. Microstructure of disperse phase of lithium grease Litol-24 (*a*, *b*) and complex lithium grease ITMOL-150 (*c*, *d*) of standard composition (*a*, *c*) and containing nano-sized UDDG additive (*b*, *d*).

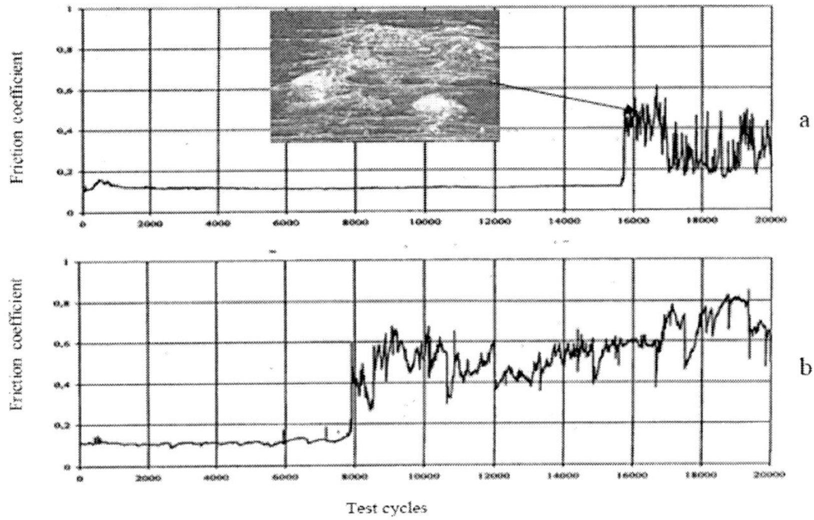

Figure 2.3. Variations in friction coefficient during testing under pressure of 20 MPa for annealed steel 45 with hardness of HB 250–270 (*a*) and hardened steel 45 with hardness of HRC 42–45 (*b*) under lubrication with grease Litol-24+UDDG prepared by technology 1 (counterbody – hardened steel 60G with hardness of HRC 58–60).

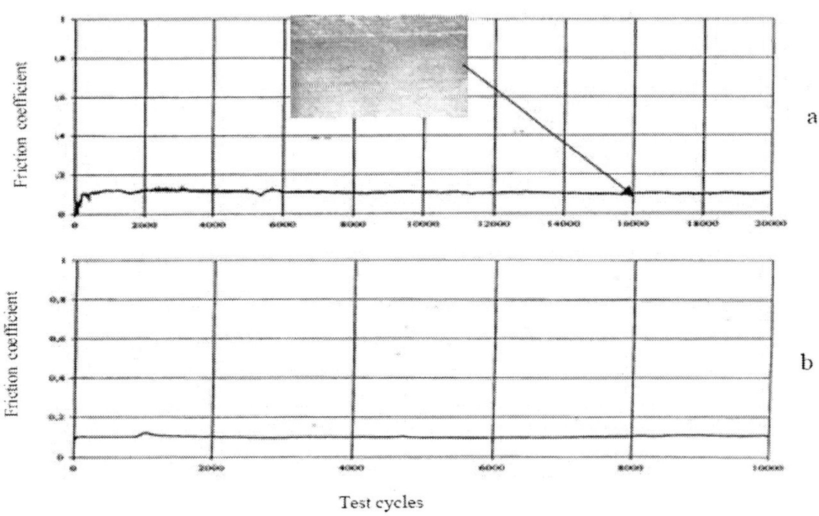

Figure 2.4. Variations in friction coefficient during testing under pressure of 20 MPa for annealed steel 45 with hardness of HB 250–270 (*a*) and hardened steel 45 with hardness of HRC 42–45 (*b*) under lubrication with grease Litol-24+UDDG prepared by technology 2 (counterbody – hardened steel 60G with hardness of HRC 58–60).

The data presented show that the use of the technology eliminating the fixation of nano-sized additives in the structure frame of the dispersed phase does not provide high antifriction properties of the grease for a long time. However, the implantation of nano-sized diamond-containing particles into the structure of the dispersed phase favors keeping high tribological properties of the grease during the whole test period.

The analysis of tribotest results has shown that the use of diamond nanoparticles as hard additives to greases raises their critical load and changes conditions of the formation of a lubricating film in the friction zone as well as reduces the wear rate of the contacting surfaces. The wear rate of the contacting surfaces is much governed by the initial material hardness and the positive effect of modification weakens as the hardness grows. In particular, when testing specimens for rolling friction in a machine SMT-1 it has been found that the modification of the grease Litol-24 by the UDDG additive reduces the wear rate of steel 45 (the chemical composition, wt.%: C – 0.47, Si – 0.27, Mn – 0.65, Cr – 0.25, Ni – 0.25, Cu – 0.25, Fe – the rest) with a hardness of 20–25 HRC approx. 30 times. The wear rate of steel ShKh-15 (the chemical composition, wt.%: C – 1.05, Si – 0.27, Mn – 0.30, Cr – 1.5, Ni – 0.30, Cu – 0.25, Fe – the rest) with a hardness of 35–40 HRC decreases five times while that of the same steel with a hardness of 62–65 HRC slightly increases, i.e. from $0.07 \cdot 10^{-7}$ to $0.08 \cdot 10^{-7}$ g/m (Table 2.2.)

Table 2.2. Tribological Characteristics of Friction Pair Materials

Characteristic	Material	Litol-24	Litol-24 + UDDG
Critical load, N		710	750
Welding load, N		1410	1600
Wear rate, g/m · 10^{-7}	Steel 45 (HRC 20–25)	4.04	0.12
	Steel ShKh-15 (HRC 35–40)	0.63	0.11
	Steel ShKh-15 (HRC 62–64)	0.07	0.08

The physical and tribological characteristics of the complex lithium grease (cLi) containing a set of additives for heavy-loaded friction units and the universal grease Litol-24 most spread in the CIS countries and the lithium grease Shell Retinax EP 2 are compared in figures 2.5 and 2.6.

The comparison of the physical and tribological characteristics of these greases shows that the complex lithium grease with the set of nano-sized additives surpasses the grease Shell Retinax EP 2 in the welding load, critical load, scoring index, and colloidal stability while their maximal working temperature, thermal hardening, and vaporability are similar. This predetermines its better antiscoring and antiwear properties.

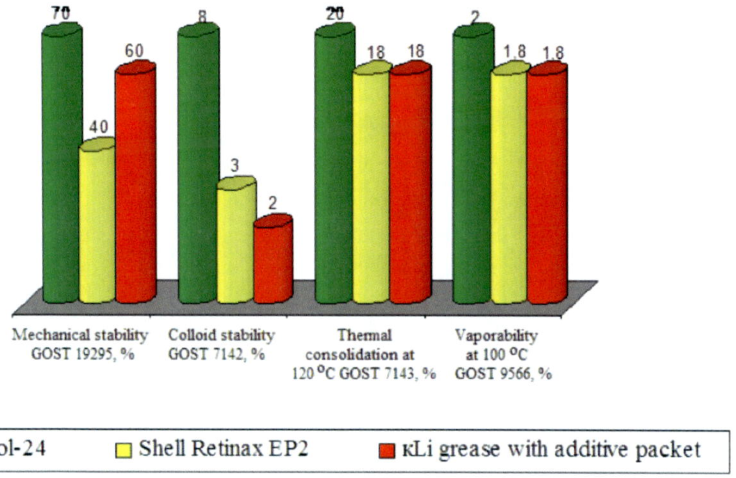

Figure 2.5. Comparison of physical characteristics of some greases.

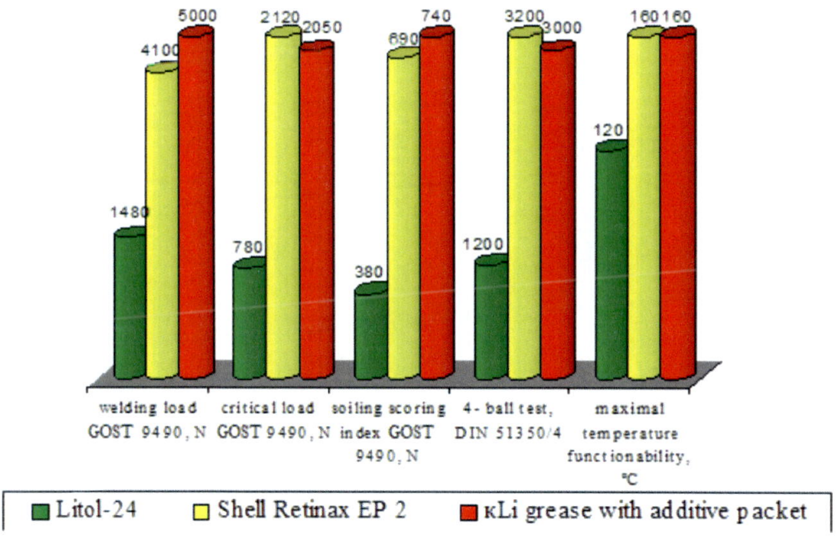

Figure 2.6. Comparison of tribological characteristics of some greases.

The distribution of the electron work function over the friction surface determined from variations in the contact voltage between a reference specimen and the studied surface (the Kelvin method) [58] and characterizing the non-uniformity of the deformation in the surface layer, hence, the non-uniformity of contact shows that in case of using the modified UDDG additive the friction surface becomes more homogeneous (figure 2.7).

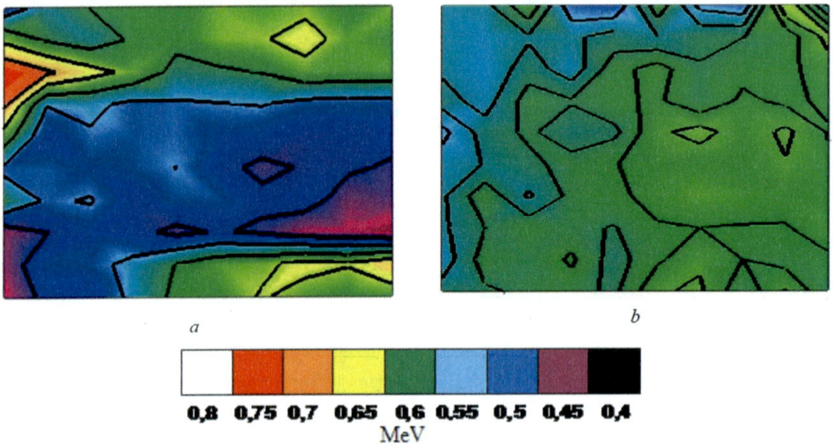

Figure 2.7. Topology of electron work function over friction surface of hardened steel 45 specimen lubricated with grease Litol-24 of standard composition (*a*) and grease Litol-24+UDDG (*b*).

This evidences that the application of the grease modified by UDDG particles favors the appearance of a more homogeneous and stable lubricating film in the friction zone compared to the application of the grease containing no hard nano-sized particles.

References

[1] Patent 5906 (BY)
[2] Patent 10897 (BY)
[3] Patent 2030449 (RU)
[4] Patent 2178803 (RU)
[5] Patent 1595885 (SU)

[6] Dolmatov, V. Yu. Detonation-Synthesized Ultradispersed Diamonds. Production, Properties, and Applications; St Petersburg State Polytechnic University: St Petersburg, RU, 2003; 344 p.
[7] Patent 2162878 (RU)
[8] Patent 2225879 (RU)
[9] Patent 2163921 (RU)
[10] Kremeshnyi, V. M., Mansurov, E. B. et al. Tribological Properties of Greases Doped with Ultradispersed Diamond-Graphite Powders; Ultradispersed Powders, Materials, and Nano-Structures: Production, Properties, and Applications; Krasnoyarsk Polytechnic Institute: Krasnoyarsk, RU, 1996; pp 190–191.
[11] Vityaz, P. A., Zhornik, V. I. et al. Tribological Properties of Friction Pairs Lubricated with Grease Modified by Ultradispersed Diamonds; Powder Metallurgy; Belarus State Scientific and Production Concern of Powder Metallurgy: Minsk, BY, 2001, Vol 24, pp 34-39.
[12] Patent 5614477 (US)
[13] Garkunov, D. N., Balabanov, V. I. *Heavy Engineering*. 2000, N 2, pp 18-23
[14] Kuzharov, A. S. Coordination Tribochemistry of Selective Transfer; Rostov-on-Don Institute of Agricultural Engineering: Rostov-on-Don, SU, 1991, 513 p.
[15] Patent WO2006/026009 (US)
[16] Patent 19949032 (DE)
[17] Patent 7102274 (JP)
[18] Patent 8470 (BY)
[19] Patent 7832 (BY)
[20] Patent 3513065 (JP)
[21] Patent 5612297 (USA)
[22] Petent 19852203 (DE)
[23] Patent 4570 (BY)
[24] Patent т 2054030 (RU)
[25] Patent 2139920 (RU)
[26] Patent 2258080 (RU)
[27] Patent 2187543 (RU)
[28] Vorobyova, S. A. *Journal of Friction and Wear*. 1996, Vol 17, 827-831.
[29] Kobylskov, I. P. Complex Solutions of Problems of Energy Saving and Reliability in Operation of Machines Based on Recovering Technology; Innovations in Machine Building; Joint Institute of Machine Building of Belarus National Academy of Sciences; Минск, BY, 2008; pp 125-128.

[30] Balabanov, V.I., Beklemyshev, V.I. et al. Friction, Wear, Lubrication, and Self-Organization in Machines; Izumrud: Moscow, RU, 2004; 192 p.
[31] Patent 2169172 (RU)
[32] Patent 2168662 (RU)
[33] Patent WO2007/082299 (США)
[34] Vityaz, P. A., Zhornik, V. I. et al. Improvement of Wear Resistance of Friction Surfaces by Running-In with Lubricant Containing Ultradispersed Additives; TRIBO-FATIGUE; Irkutsk State University of Communications: Irkutsk, RU, 2005, Vol 2, pp 80-87.
[35] Zhornik, V. I., Ivakhnik, A. V. Effect of Nano-Sized Carbon Additives on Structure of Greases and Wear Resistance of Friction Surfaces; Carbon Nano-Structures; A. V. Lykov Heat and Mass Transfer Institute of National Academy of Sciences of Belarus: Minsk, BY, 2006; pp 81-87.
[36] Rychkov, Yu. M. Applied Electrodynamics; Grodno State University: Grodno, BY, 1999; 170 p.
[37] Bereznyakov, A. I. *Journal of Friction and Wear.* 1996, Vol 17, N 1, 43-49.
[38] Savkin, V. G., Chmykhova, T. G. et al. *Journal of Friction and Wear.* 2001, Vol 22, N 5, 561-566.
[39] Kutkov, A. A. Wear-Resistant and Antifriction Coatings; Mashinostroenie: Moscow, SU, 1976; 152 p.
[40] Bershadskii, L. I. *Journal of Friction and Wear.* 1992, Vol 13, 1077-1080.
[41] Kuzharov, A. S., Akhverdiev, K. S. et al. *Journal of Friction and Wear.* 2001, Vol 22, N 1, 84-91.
[42] Ventsel, K. S. *Journal of Friction and Wear.* 1992. Vol 13, 908-916.
[43] Sosulina, L. N., Skryabina, T. G. *Journal of Friction and Wear.* 1984, Vol 5, 923-929.
[44] Ventsel, S. V., Bazderkin, V. A. et al. *Journal of Friction and Wear.* 1986, Vol 7, 301-307.
[45] Struk, V. A., Rogachev, A. V. et al. *Materials, Technologies, Tools.* 2002, Vol 7, N 3, 53-65.
[46] Dolmatov, V. Yu. *Progress in Chemistry.* 2007, N 4, 375-397.
[47] Vityaz, P. A., Zhornik, V. I., Kukareko, V. A. *Physical Mesomechanics.* 2006, Vol 9, N 5, 85-89.
[48] Tochilnikov, D. G., Ginzburg, B. M. *Journal of Technical Physics.* 1999, Vol 69, N 6, 102-105.

[49] Vityaz, P. A., Zhornik, V. I. et al. *Journal of Friction and Wear*. 2000, Vol 21, 527-533.

[50] Feklisova, T. G., Vasiliev, I. A. et al. *Journal of Friction and Wear*. 1982, Vol 3, N 2, 225-233.

[51] Lyuty, M., Kostyukovich, G. A. et al. *Journal of Friction and Wear*. 2002, Vol 23, N 4, 411-424.

[52] Bogatyreva, G. P., Voloshin, M. N. *Superhard Materials*. 1998, N 4, 82-87.

[53] Komarov, V. F., Sakovich, G. V. Properties of Nano-Sized Synthetic Diamonds as Result of Nano-Sized State of Their Particles; Detonation Nanodiamonds: Production, Properties, and Applications; A. I. Ioffe Physical-Technical Institute: St Petersburg, RU, 2003; pp 68-69.

[54] Dolmatov, V. Yu., Fujimura, T. *Superhard Materials*. 2001, N 6, 34-41.

[55] Aleksenskii, A. E., Baidakova, M. V. et al. *Physics of Solids*. 1999, Vol 41, N 4, 740-743.

[56] Chekan, V. A., Markova, L. V. et al. *Industrial Laboratory*. 2005, N 10, 19-21.

[57] Zhornik, B. I., Ivakhnik, A. V. et al. Structure and Properties of Greases Modified by Nano-Sized Diamond-Graphite Additives; Machine-Building and Technosphere of the 21th Century; Donetsk National Technical University: Donetsk, UA, 2006; Vol 2, pp 6-41.

[58] Kasai, T., Fu, X.Y., Rigney, D.A., Zharin, A.L. Application of a Non-Contacting Kelvin Probe during Sliding / *Proc. Int. Conf. Wear of Materials,* Atlanta, April 1999 and Wear 225-229 (1999) 1186-1204.

Chapter 3

Tribological Properties of Materials and Coatings Lubricated by Lubricants with Nanodiamonds

3.1. Tribological Properties of Polymer Composites Lubricated by Grease with Nanodiamonds

Antifriction polymer composites (APC) belong to a vast group of advanced materials finding increasing application in various branches of machine-building. The self-lubricating properties of some polymer matrices and, chiefly, good antifriction properties of many reinforcing fillers in combination with solid lubricants have resulted in the development of antifriction polymer composites that possess high strength and tribological characteristics. They are efficient under heavy pressures (up to 30–40 MPa) [1–3].

To estimate the influence of nano-sized hard additives to lubricants on the tribological behavior of polymer materials and to develop recommendations on the application of modified greases in polymer composite friction units, different variants of APC were tested with the common lithium grease Litol-24 and the complex lithium grease ITMOL-150 containing the UDDG additive.

Specimens of the following materials were tested: the three-layer metal-polymer composites (MPC) DX1 and FR.EX (Glacier Garlock Bearings, Slovakia), the polymer composite (PC) Oksafen (JSC “SRC “Viscose”, Russia), the polymer composite based on phenol resin with synthetic fibers and graphite IMMS-A developed at the V. A. Belyi Metal-Polymer Research

Institute of the National Academy of Sciences of Belarus, and the polymer composite based on epoxy resin with synthetic fibers and nano-sized particles IGNCH-1 developed at the Institute of General and Inorganic Chemistry of the National Academy of Sciences of Belarus. The tests were performed in a friction machine MTBP-2 involving the reciprocal motion of a specimen over the counterbody under the nominal pressure of 35, 50, and 75 MPa. The maximal velocity of reciprocal motion was ≈ 0.2 m/s. The counterbody was made of plates of heat treated steel 40XH with the hardness of 52–54 HRC and the following chemical composition, wt.%: C – 0.40, Si – 0.27, Mn – 0.65, Cr – 0.60, Ni – 1.20, Cu – 0.30, S – 0.035, P – 0.035, Fe – the rest). The roughness of the counterface was $R_a = 0.32$ μm.

Figure 3.1 shows the images of the cross section and surface of the metal-polymer composite DX1 before tests and after one of test stages.

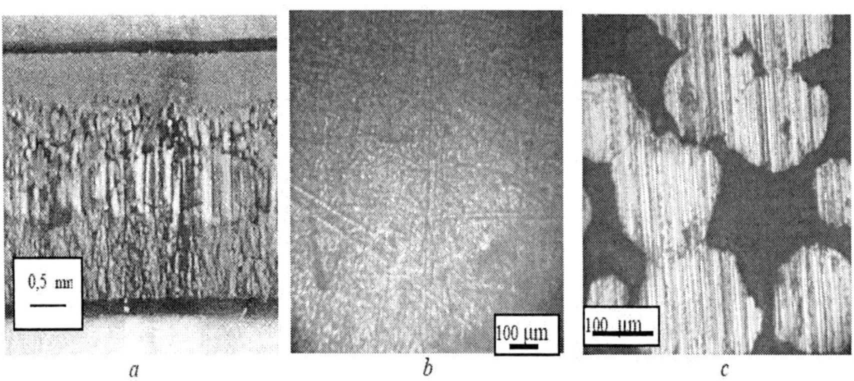

Figure 3.1. Microstructure of cross section of metal-polymer composite DX1 (*a*) and friction surface in original state (*b*) and after passing sliding distance of 2.2 km (*c*).

During running-in the friction coefficient of the metal-polymer composites DX1 and FR.EX is 0.08–0.09 while at the steady-state wear it decreases to 0.06–0.07. As the thickness of the polymer layer decreases (the sliding distance is 1–5 km), bronze particles appear on the friction surface. After a sliding distance of ≈ 5 km for the composite DX1 and ≈ 6 km for the composite FR.EX the bronze substrate with polymer inclusions outcrops (figure 3.2).

The friction coefficient remains at the level of 0.07. The microhardness of the bronze substrate on the friction surface of the composite DX1 is $H_{0.98}$ = 2200 MPa while on the surface of the composite FR.EX it is $H_{0.98}$ = 2150 MPa. The friction coefficients of the composites Oksafen, IMMS-A, and

IGNCH-1 are similar and amount to 0.07–0.08. The Vickers hardness is HV_{10} = 260 MPa for the composite Oksafen, HV_{10} = 290 MPa for the composite IMMS-A, and HV_{10} = 480 MPa for the composite IGNKH-1.

a *b*

Figure 3.2. Image of friction surface (*a*) and microstructure (*b*) of specimen of metal-polymer composite DX1 after passing sliding distance of 6.3 km under pressure of 75 MPa.

Figure 3.3 shows data on the linear wear of the metal-polymer composites DX1 and FR.EX as well as the polymer composites Oksafen, IMMS-A, and IGNCH-1 lubricated with the greases Litol-24 and ITMOL-150N+UDDG under pressures of 25, 50, and 75 MPa.

The effect of grease replacement is the most pronounced for the MPC FR.EX which shows relatively quick failure of the polymer layer when using the commercial grease Litol-24, especially under pressures of 50 and 75 MPa (figure 3.3, *b*). The application of the grease ITMOL-150N+UDDG favors a considerable deceleration of this process (1.45–2.42 times) that is confirmed by the images of the working surfaces of specimens shown in figure 3.4. As it is seen in the images, traces of the polymer film present on the surface of the FR.EX specimen after passing the sliding distance of 6.2 km with the grease ITMOL-150N+UDDG while the surface of the same specimen lubricated with the grease Litol-24 shows no polymer film already after passing the distance of 3 km (under a pressure of 50 MPa).

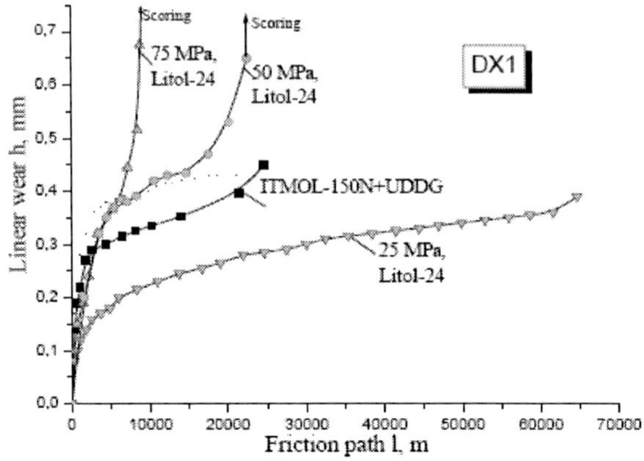
Scoring
Scoring
75 MPa, Litol-24
50 MPa, Litol-24
DX1
ITMOL-150N+UDDG
25 MPa, Litol-24
Linear wear h, mm
Friction path l, m

a

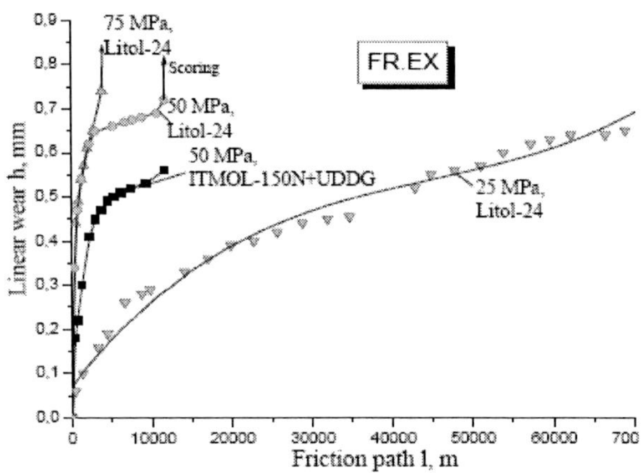
75 MPa, Litol-24
Scoring
FR.EX
50 MPa, Litol-24
50 MPa, ITMOL-150N+UDDG
25 MPa, Litol-24
Linear wear h, mm
Friction path l, m

b

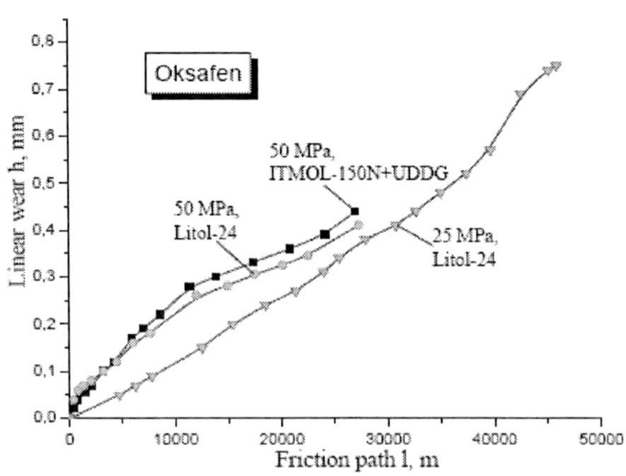
Oksafen
50 MPa, ITMOL-150N+UDDG
50 MPa, Litol-24
25 MPa, Litol-24
Linear wear h, mm
Friction path l, m

c

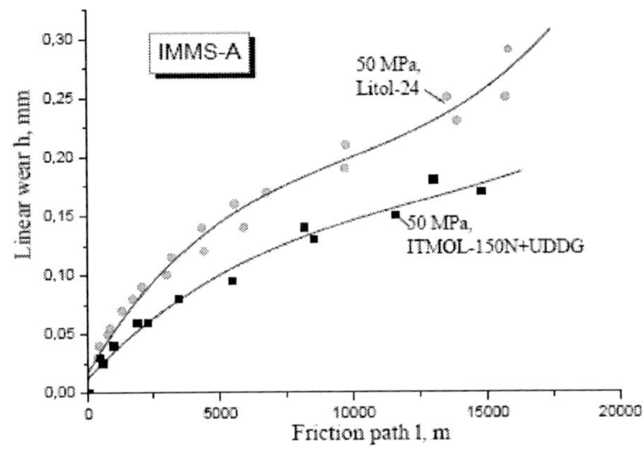

d

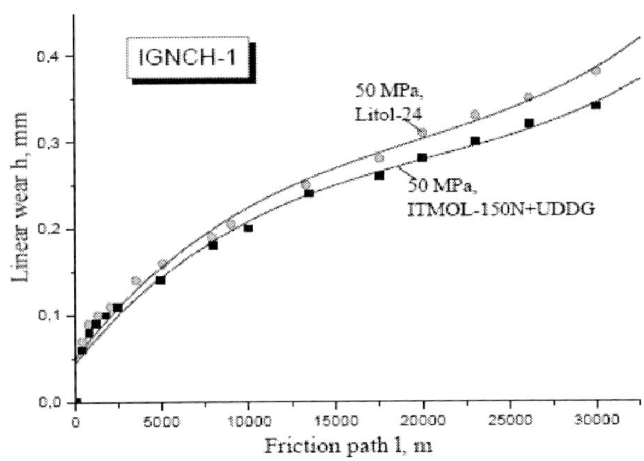

e

Figure 3.3. Dependence of linear wear of specimens of polymer composites lubricated with greases Litol-24 and ITMOL-150N+UGGD under various loads on sliding distance.

A similar effect of decrease in the wear rate occurs also when lubricating a DX1 specimen with ITMOL-150N+UDDG though it is less pronounced (1.41–2.03 times) (figure 3.3, *a*). This can probably be explained by the fact that the metal-polymer composite DX1 has a higher initial hardness and wear resistance compared to the composite FR.EX. The general view of the specimens of the metal-polymer composite DX1 tested with the greases Litol-24 and ITMOL-150N+UDDG after passing the sliding distance of about 6.5 km is shown in figure 3.5.

a *b* *c*

Figure 3.4. View of friction surface of FR.EX specimen after passing sliding distance of 3 km in grease Litol-24 (*a*) and after passing sliding distance of 3 km (*b*) and 6.2 km (*c*) in grease ITMOL-150N+UDDG under pressure of 50 MPa.

a *b*

Figure 3.5. View of friction surface of DX1 specimen after passing sliding distance of 6.5 km in greases Litol-24 (*a*) and ITMOL-150N+UDDG (*b*).

Specimens of the polymer composites Oksafen, IMMS-A, and IGNCH-1 show the best wear resistance under a pressure of 50 MPa. The duration of their running-in is quite short and the linear stage of steady-state friction begins already after passing the distance of 0.5–1.0 km. The wear rate at this stage of friction is $I_h = (0.9–1.1)\cdot 10^{-8}$ that exceeds slightly the corresponding values of the wear rate for the metal-polymer composites. However, the friction of the pairs Oksafen – steel 40X, IMMS-A – steel 40X, and IGNCH-1 – steel 40X does not turn to adhesive interaction because of the absence of metallic contact and the retarded degradation of the grease. For this reason the life of the polymer composites Oksafen, IMMS-A, and IGNCH-1 exceeds considerably that of the metal-polymer linings FR.EX and DX1.

The data presented in figure 3.3 show that the polymer composites Oksafen, IMMS-A, and IGNCH-1 have the lowest linear wear. For example, under a pressure of 50 MPa the average wear rate of the composites Oksafen,

IMMS-A, and IGNCH-1 lubricated with Litol-24 is $I_h = 1.42{\cdot}10^{-8}$, $I_h = 1.36{\cdot}10^{-8}$, and $I_h = 1.60{\cdot}10^{-8}$, respectively.

Figure 3.6 shows the images of the friction surface of specimens of the polymer composites after testing with the grease Litol-24. It is seen that the surfaces of the composites having a high wear resistance are smooth and contain no signs of delamination and chipping.

Figure 3.6. View of friction surfaces of polymer composites Oksafen (*a*), IMMS-A (*b*), and IGNCH-1 (*c*) after tribotests in grease ITMOL-150N under pressure of 50 MPa.

The presented data on the wear of the surface of the metal-polymer composites show that the use of the grease modified by ultradispersed hard particles of various origin within the pressure range 25–50 MPa reduces the material wear rate; the efficiency of the additives increases with increasing the contact pressure. Under a pressure of 75 MPa the positive effect of using the modified grease occurs only when it is applied at the initial stage of testing, i.e. during running-in.

The improvement of the wear resistance of the metal-polymer composites lubricated with a grease containing ultradispersed hard additives can be explained by the joint effect of the factors favoring increase in tribological properties of the grease and changing the structure of the friction surface owing to precipitation hardening and tribomechanical modification. Nanoparticles contained in the grease dispersed phase raise the colloidal stability of the grease and the oil-retaining capacity of its structure frame as well as give the grease antiscoring properties. The provision of a regular lubricating regime in the contact zone favors decrease in the thermal load of the metal-polymer composite and increase in its structure stability. Hard nanoparticles contained in the grease form a wear-resistant cell structure of the metallic surface owing to severe plastic deformation and provide the precipitation hardening of the contacting surfaces due to their embedding in friction [4].

Decrease in the wear rate of the polymer composites Oksafen, IMMS-A, and IGNCH-1 can be explained by the growth of the carrying capacity of the grease with nano-sized particles. In transition from the grease Litol-24 to the modified grease ITMOL-150N+UDDG the wear resistance of the composite Oksafen diminishes that may likely be caused by peculiarities of the structure of its binder and requires a more thorough study. The predictive estimate of the life of friction pairs with different material combinations as applied to the insert of the spherical joint of a mounting cylinder of a BelAZ rock handler can be performed by comparing it with that of the base variant of the friction pair proceeding from the wear rate averaged over the test period (Table 3.1).

Table 3.1. Predictive Life of Various Friction Pairs Found from Test Results

Specimen material	Grease	Pressure, MPa	Average wear rate, μm/km	Permissible wear, mm	Predictive life, km
MPC DX1	Litol-24	25	5.9	0.4	67.8
		50	26.2		15.3
	ITMOL-150H	25	2.9		137.9
		50	18.5		21.6
MPC FR.EX	Litol-24	25	9.3		43.0
		50	65.2		6.1
	ITMOL-150H	25	3.84		104.1
		50	44.9		8.9
PC Oksafen	Litol-24	25	3.4	1.0	294.1
		50	14.2		70.4
	ITMOL-150H	25	3.7		108.1
		50	16.2		61.7
PC IMMS-A	Litol-24	25	3.2		312.5
		50	16.0		62.5
	ITMOL-150H	25	2.6		384.6
		50	10.6		94.3
PC IGNCH-1	Litol-24	25	3.0		333.3
		50	13.6		73.5
	ITMOL-150H	25	2.5		400.0
		50	10.4		96.1

The friction pair MPC DX1 – steel 40XH was selected as the base variant and the data obtained under pressures of 25 and 50 MPa were used for comparison.

The analysis of these data shows that three variants of the composition of PC have similar wear rate and predictive life of the friction unit. In particular, under a pressure of 50 MPa the average wear rate of the materials Oksafen, IMMS-A, and IGNKH-1 lubricated with the grease Litol-24 is $I_h = 1.42 \cdot 10^{-8}$, $I_h = 1.36 \cdot 10^{-8}$, and $I_h = 1.60 \cdot 10^{-8}$, respectively. This provides the predictive life of 70.4 km for the composite Oksafen, 73.5 km for the composite IGNCH-1 and 62.5 km for the composite IMMS-A while the predictive life of the metal-polymer lining DX1 is 15.3 km. In this case the application of the grease with nano-sized hard additives provides decrease in the wear rate of the metal-polymer linings 1.4–2.4 times while that of the PC IMMS-A and IGNCH-1 – 1.2–1.5 times. This is explained by the growth of the carrying capacity of the modified grease layer as well as by the hardening of the metallic component of the metal-polymer composites. With account for the augmentation of the permissible linear wear from 0.4 to 1.0 mm in transition from the three-layer metal-polymer composites to the polymer composites and having in mind a lower wear rate of the latter, the predictive life of the insert of the spherical joint of a mounting cylinder of a BelAZ rock handler can be prolonged 5.9–6.2 times when using the studied polymer composites.

Results of the study of the tribological behavior of the polymer composites have shown that three variants of their compositions (Oksafen, IGNCH-1, and IMMS-A) have similar wear rates and provide similar predicted lifes of a friction unit. In particular, under a pressure of 50 MPa the average wear rate of the materials Oksafen, IMMS-A, and IGNKH-1 lubricated with the grease Litol-24 is $I_h = 1.42 \cdot 10^{-8}$, $I_h = 1.36 \cdot 10^{-8}$, and $I_h = 1.60 \cdot 10^{-8}$, respectively. This provides the predictive life of 70.4 km for the composite Oksafen, 73.5 km for the composite IGNCH-1 and 62.5 km for the composite IMMS-A while the predictive life of the metal-polymer lining DX1 is 15.3 km. In this case the application of the grease with nano-sized hard additives provides decrease in the wear rate of the metal-polymer linings 1.4–2.4 times while that of the PC IMMS-A and IGNCH-1 – 1.2–1.5 times. This is explained by the growth of the carrying capacity of the modified grease layer as well as by the hardening of the metallic component of the metal-polymer composites.

The application of the polymer composites IMMS-A and IGNCH-1 to fabricate the insert of the spherical joint of a mounting cylinder of a BelAZ rock handler instead of the MPC DX-1 and the use of the grease containing nano-sized additives will allow prolonging the unit life 5.9–6.2 times.

3.2. Tribological Properties of Metallic Materials Lubricated by Grease with Nanodiamonds

We have studied the effect of the pressure on the tribological behavior of the pairs steel 45 – steel 45 and copper M1 – steel 45 lubricated with commercial grease Litol-24 (State Standard GOST 21150-75) and the same grease modified by the additive of ultradispersed diamond-graphite charge (UDDG) (Technical Specifications of Belarus 28619110.001-95) in an amount of 1 wt.%. The reciprocal sliding tests were carried out at an average velocity of relative movement of ≈ 0.1 m/s. The wear was determined from the mass loss over a sliding distance of 3–5 km.

The test results are presented in figures 3.7 and 3.8. They show that the influence of the modified additive to the grease on wear processes depends substantially on the pressure. Under a relative light nominal contact pressure (p_a = 10 MPa) the wear of steel 45 specimens lubricated with the modified grease exceeds considerably (approx. twice) the wear of specimens lubricated with the commercial grease (figure 3.7, a). At initial stages of testing the friction coefficient of the pair lubricated with the modified grease is f = 0.10–0.11 (figure 3.7, b). Subsequently, as the sliding distance increases, the friction coefficient gradually decreases down to f = 0.08–0.09 at $l \approx$ 1.0–1.5 km and f = 0.07–0.08 at $l \approx$ 2.5–3.0 km. These changes in the friction coefficient with increasing the sliding distance are caused by the running-in of the friction pair.

When using the non-modified grease, the pattern of the dependence of the friction coefficient on the sliding distance is similar to that for the grease containing UDDG. However, the running-in of the contacting surfaces in the non-modified grease is slowed and the friction coefficient at late stages of testing exceeds the friction coefficient for the grease containing UDDG. The relatively slowed kinetics of running-in in the non-modified grease is also confirmed by a lower mass wear of steel.

Increase in the nominal pressure up to p_a = 20 MPa makes the dependences of the mass wear of steel on the sliding distance for the modified and non-modified greases closer (figure 3.7, c). We note, however, that after passing a sliding distance of 2.5 km the wear rate of the specimens drops sharply. The friction coefficient at this testing stage also decreases considerably (f = 0.060–0.065) (figure 3.7, d).

The wear rate of steel lubricated with the non-modified grease Litol-24 remains high during the testing, while the friction coefficient at late testing stages even somewhat increases reaching *f* = 0.110–0.115.

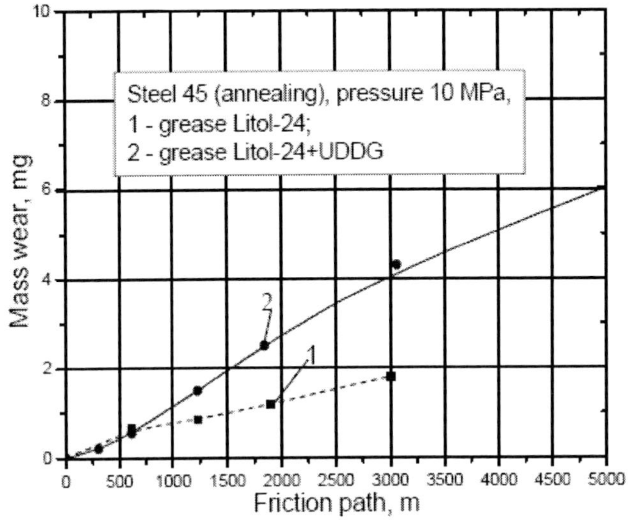

a

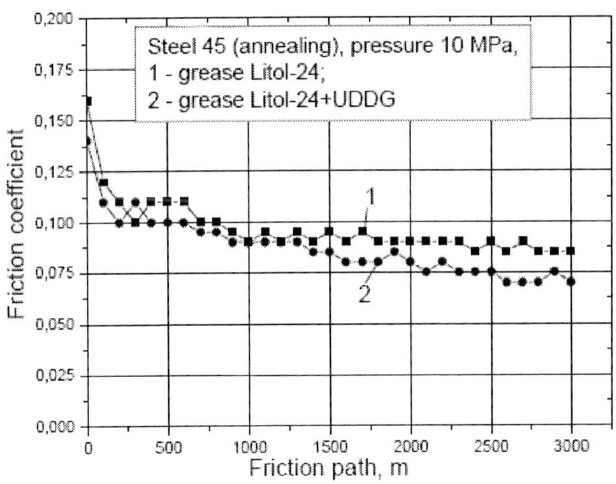

b

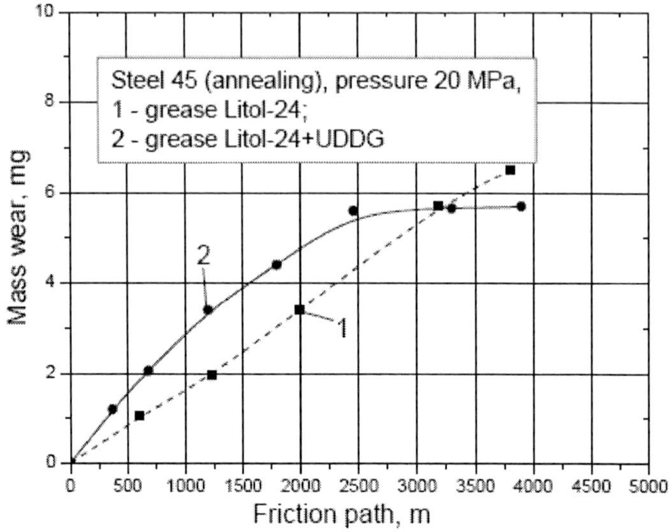
Steel 45 (annealing), pressure 20 MPa,
1 - grease Litol-24;
2 - grease Litol-24+UDDG
Mass wear, mg
Friction path, m
0
2
4
6
8
10
500
1000
1500
2000
2500
3000
3500
4000
4500
5000
1
2

c

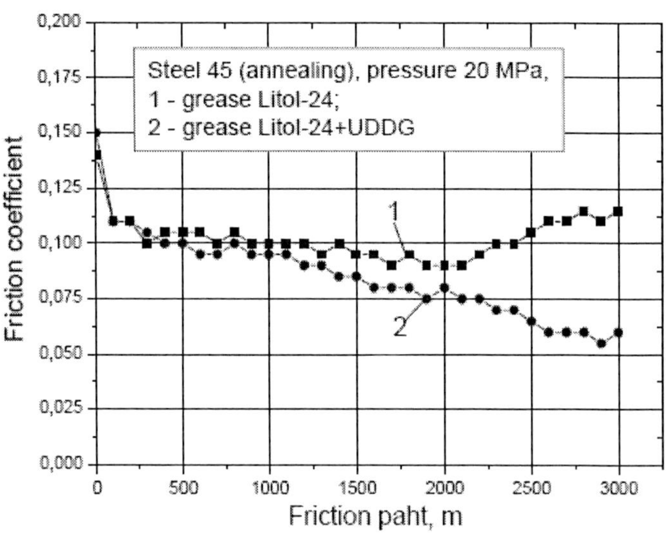
Steel 45 (annealing), pressure 20 MPa,
1 - grease Litol-24;
2 - grease Litol-24+UDDG
Friction coefficient
Friction paht, m
0,000
0,025
0,050
0,075
0,100
0,125
0,150
0,175
0,200
0
500
1000
1500
2000
2500
3000
1
2

d

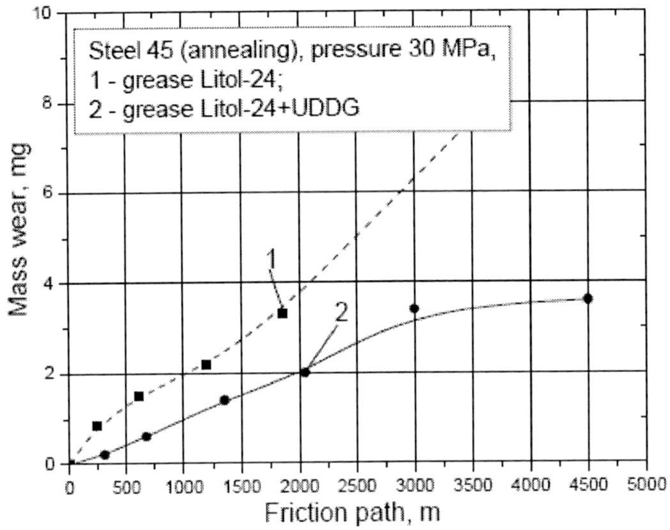

e

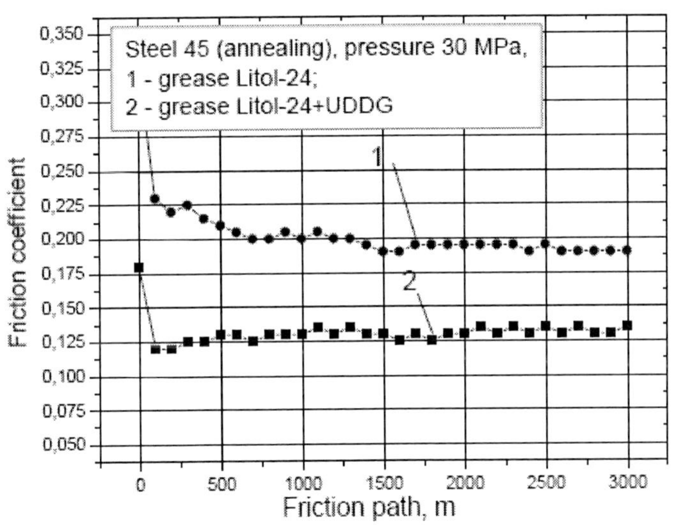

f

Figure 3.7. Dependences of mass wear of annealed steel (*a*, *c*, *e*) and friction coefficient (*b*, *d*, *f*) on sliding distance under various pressures: *a*, *b* – 10 MPa; *c*, *d* – 20 MPa; *e*, *f* – 30 MPa.

Under a contact pressure of p_a = 30 MPa, the use of the grease modified by UDDG results in a considerable drop of the wear rate of the friction surface compared to lubrication with the non-modified grease (figure 3.7, e). The friction coefficient decreases down to f = 0.060–0.065 already at early testing stages (figure 3.7, f). The tests in the non-modified grease under high contact pressures (p_a = 30 MPa) are accompanied by wear acceleration and the rise of the friction coefficient up to f = 0.12–0.13 already at the very beginning of testing.

The effect of the pressure on the running-in and tribological behavior of the pair annealed steel 45 – hardened steel 65G was studied in grease Litol-24 under nominal contact pressures of 10, 20, and 30 MPa. The results have shown that with increasing the pressure the friction coefficient and wear rate in the modified grease become decreasing at earlier stages of running-in. One of the possible reasons of this could be the hardening of surface layers of steel due to their modification by ultradispersed diamond particles during friction.

To confirm this assumption we carried out additional tribotests with the non-modified grease Litol-24; the specimens were preliminarily run-in in the modified grease. The test results show that the specimens treated in such a way retain a high wear resistance and a low friction coefficient (figure 3.8).

The study of the influence of the pressure on the tribological behavior of the friction pairs has shown that under boundary lubrication of the steel – steel pair the modification of greases by additives of ultradispersed diamond-graphite charge is efficient under pressures over 20 MPa. It is enough to apply the modified grease only during running-in since after it a pair lubricated with the non-modified grease demonstrates a quite low friction coefficient (f = 0.04–0.05) and almost zero wear.

The authors of [5] studied the dependence of the wear of steel 45 hardened under various conditions on the test duration. The tests were carried out in grease Litol-24 both commercial and modified by UDDG additives under a pressure of 30 MPa (figure 3.9). These data show that the introduction of diamond-graphite charge into the grease accelerates wear and increases the friction coefficient of steel 45 hardened under various conditions. In this case the wear of the pair during running-in increases and transition from steady to catastrophic wear accelerates. It is of interest that increase in the hardness of steel 45 reduces the wear resistance of the pair in the grease modified by UDDG.

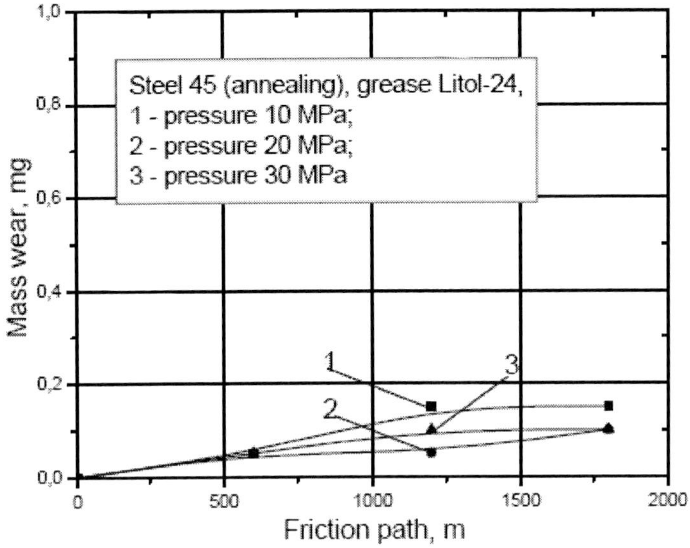

a

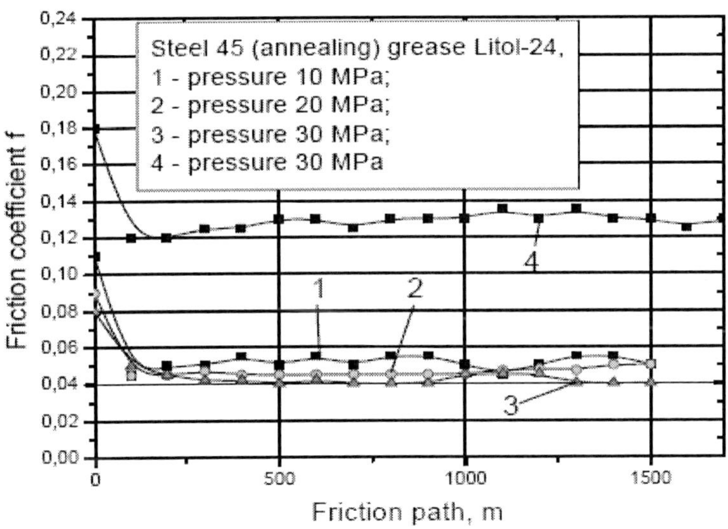

b

Figure 3.8. Dependences of mass wear (*a*) and friction coefficient (*b*) of annealed steel 45 on sliding distance after running-in under various pressures: *1*, *2*, *3* – lubrication with grease Litol-24 modified by UDDG; *4* – lubrication with commercial grease Litol-24.

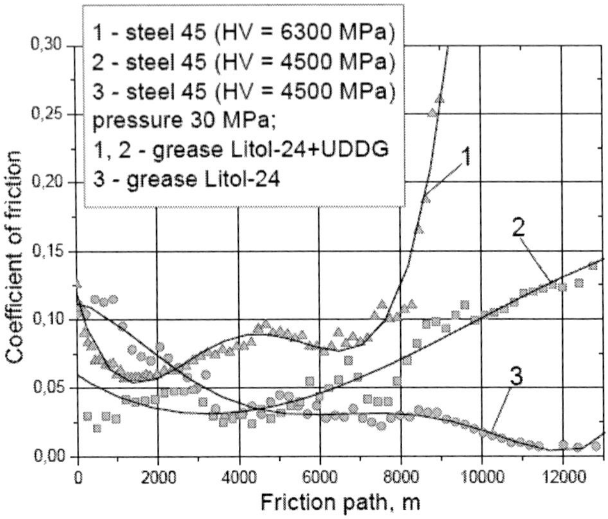

a

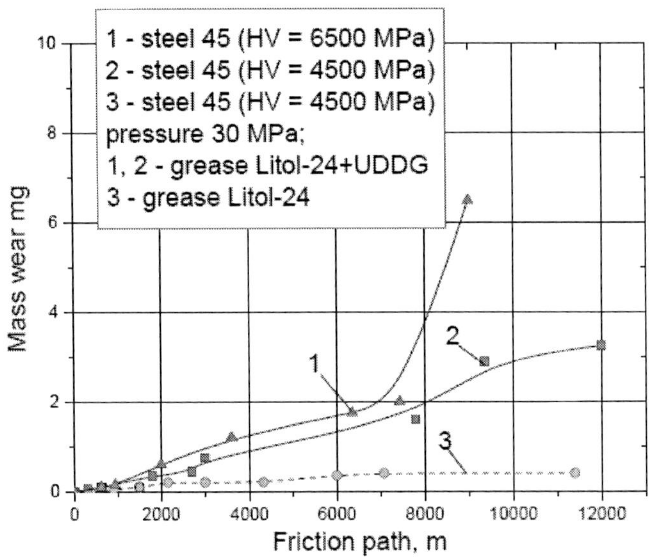

b

Figure 3.9. Dependences of mass wear (*a*) and friction coefficient (*b*) of steel 45 hardened under various conditions on sliding distance under pressure of 30 MPa: *1* – *HV* = 6500 MPa; *2*, *3* – *HV* = 4500 MPa; *1*, *2* – grease Litol-24+UDDG; *3* – grease Litol-24.

We studied the influence of regimes of the friction of the pair copper M1 – hardened steel 45 lubricated with grease Litol-24 on properties of the friction surfaces (copper M1 had the following chemical composition, wt.%: Cu – no less than 99.9, Bi – 0.001, Sb – 0.002, As – 0.002, Fe – 0.005, Ni – 0.002, Pb – 0.005, Sn – 0.002, S – 0.005, and Zn – 0.005). The results have shown that the influence of the UDDG modified additive also depends substantially on the friction regimes including the pressure. Under a relatively light nominal contact pressure (p_a = 5 MPa), the friction coefficient of the pair lubricated with the modified grease is f = 0.11–0.12 at the beginning of testing. Subsequently, it gradually decreases down to f = 0.07–0.08 at $l \approx$ 500–750 m and f = 0.065–0.070 at $l \approx$ 1.0–1.5 km (figure 3.10). For the non-modified grease the pattern of the dependence of the friction coefficient on the sliding distance is similar to that described above. The friction coefficient at late stages of testing (f = 0.085–0.090 at $l \approx$ 1.0–1.5 km) exceeds the friction coefficient for the grease containing the UDDG additives. The wear rate of copper M1 in the modified grease Litol-24 is $I = (0.25–0.30)\cdot10^{-8}$ that is 1.3–1.6 times higher than that of specimens tested in the non-modified grease.

With increasing the pressure up to p_a = 10 MPa tribological processes in the modified grease intensify that appears as decrease in the friction coefficient down to f = 0.050–0.055 over a sliding distance of $l \approx$ 750–800 m. For the non-modified grease the friction coefficient at late stages of testing is f = 0.10–0.11. The wear rates of copper M1 in the modified and non-modified greases become closer and reach $I = (0.20–0.25)\cdot10^{-8}$.

Under a pressure of p_a = 20 MPa, the modification of grease Litol-24 by UDDG additives reduces the wear rate of copper down to $I = (0.10–0.15)\cdot10^{-8}$ and the friction coefficient down to f = 0.040–0.055 after a short period of testing ($l \approx$ 200–250 m). When using the non-modified grease under heavy pressures (p_a = 20 MPa), wear intensifies ($I = (0.75–0.80)\cdot10^{-8}$) and the friction coefficient augments up to f = 0.11–0.12 during almost the whole test cycle.

The stabilization of the friction coefficient and wear rate after passing a sliding distance of 600–800 m retained up to passing a distance of 3–5 km. The tribotests revealed the dependence of the modifying effect of the ultradispersed diamond-graphite additives on the rheological behavior of greases. We tested the hardened steel 45 – cast iron KCh 30-6 pair lubricated with the non-modified grease Litol-24 under a pressure of p_a = 10 MPa at a sliding velocity of 0.6 m/s. Cast iron had the following chemical composition, wt.%: C – 2.6, Si – 1.1, Mn – 1.0, S – 0.1, P – 0.2, and the rest was Fe. It has been found that with elevating the temperature from 20 to 140 °C the friction coefficient changed from 0.04 to 0.13 and the pattern of the changes was similar to the pattern of temperature elevation (figure 3.11, a, c).

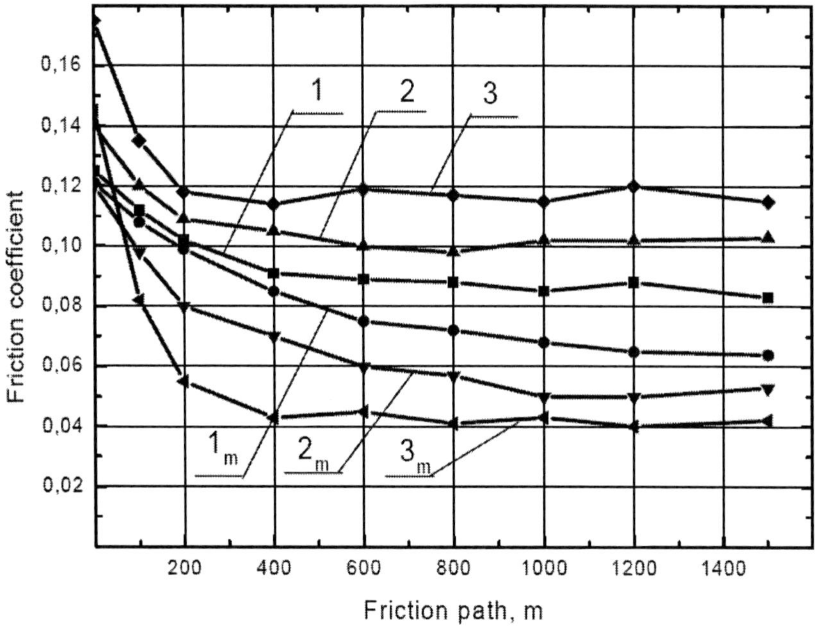

Figure 3.10. Variations in friction coefficient of copper M1 – steel 45 pair lubricated with commercial grease Litol-24 (*1*, *2*, *3*) and grease Litol-24 modified by UDDG (1_m, 2_m, 3_m) under various pressures: *1*, 1_m– 5 MPa; *2*, 2_m – 10 MPa; *3*, 3_m – 20 MPa.

However, when using the grease Litol-24 modified by UDDG, the friction coefficient was stabilized at a level of f = 0.09–0.10 within a broad temperature range (60–100 °C) (figure 3.11, b, d). In this case the negative effect of decrease in the grease viscosity due to heating becomes weaker. For the modified grease a higher friction coefficient at initial stages of testing (at temperatures below 50 °C) can be explained by the more intensive running-in of the rubbing surfaces. The wear rate of specimens was I = 0.032 mg/km for the non-modified grease Litol-24 and I = 0.026 mg/km for the modified grease.

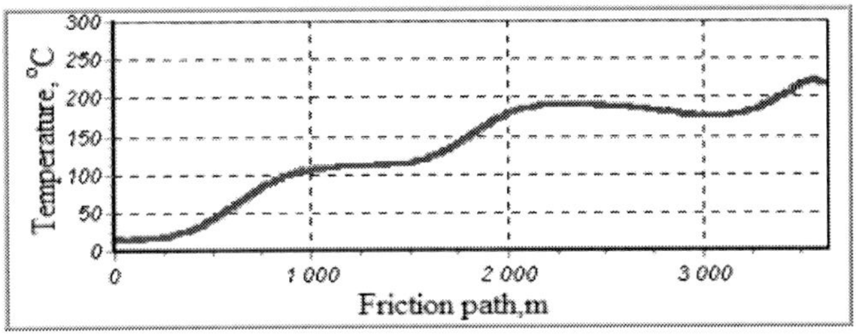

a

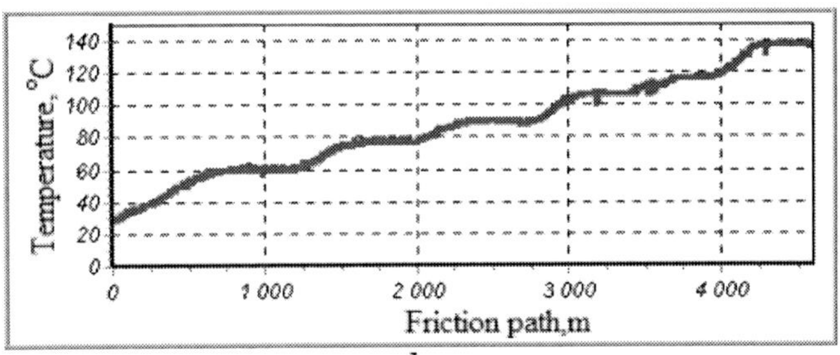

b

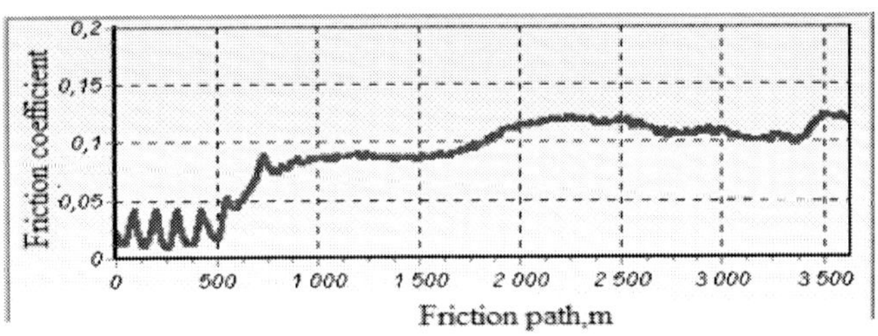

c

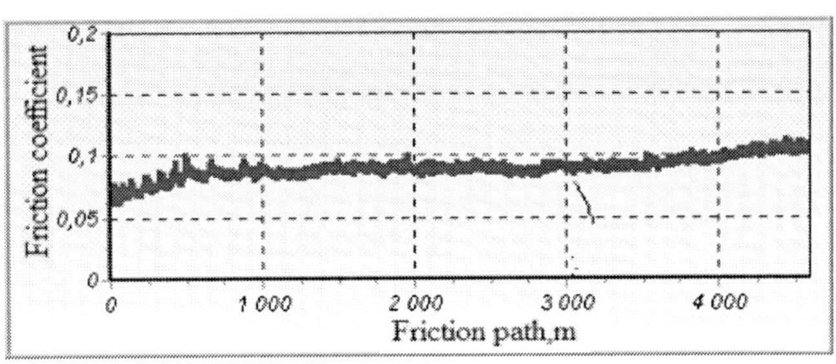

d

Figure 3.11. Variations in testing temperature (*a*, *b*) and friction coefficient (*c*, *d*) of steel 45 – cast iron KCh 30-6 pair lubricated with commercial (*a*, *c*) and UDDG-modified (*b*, *d*) grease Litol-24.

3.3. Tribological Properties of Gas-Thermal Coatings Lubricated by Grease with Nanodiamonds

Friction units of technological and transport machinery as well as machine-tools comprise parts made, as a rule, of steels, cast irons, nonferrous metals, and alloys. Practice has shown that their recovery is most efficient from technical and economical viewpoints if the technologies of the deposition of wear-resistant and antifriction coatings are used. These technologies involve the gas-thermal spraying (GTS) of wire materials (figure 3.12), i.e., the gas-flame spraying (GFS) of wires [6], electric arc metallization (EAM) [7], and hypersonic electrometallization (HM). The HM method has been developed in Belarus [8] and combines the advantages of the EAM process (simple implementation and high output) with those of high-speed gas-flame spraying (the HVOF process). To increase the speed of particles of the wires being sprayed HM set-ups produce a jet of combustion products of propane-air mixture which is effused from a special chamber at a hypersonic speed (1200–1500 m/s).

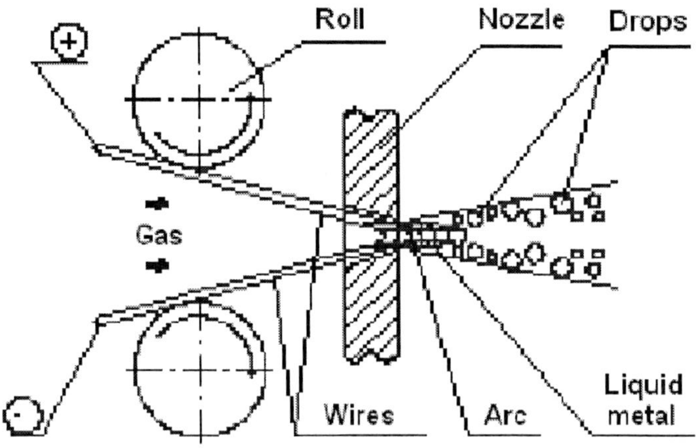

a

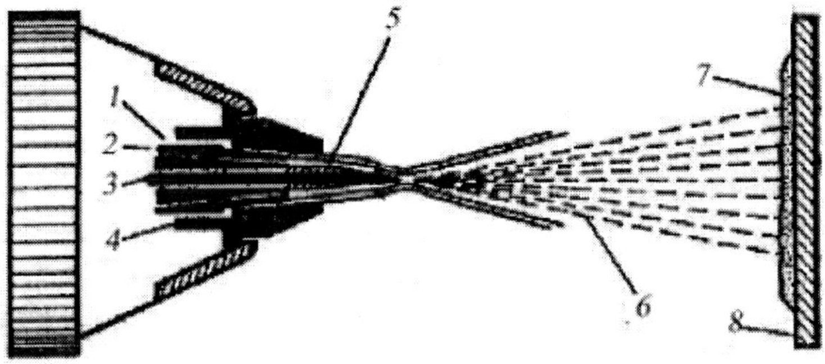

b

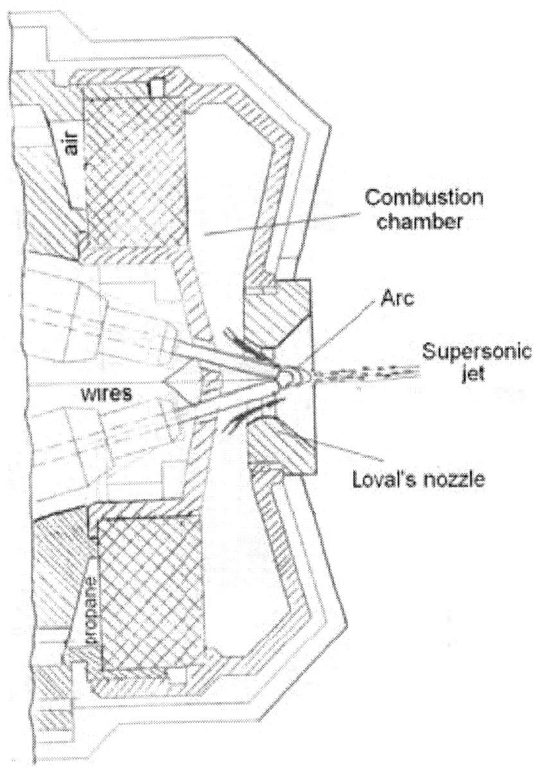

c

Figure 3.12. Schematic diagrams of coating deposition by wire spraying: *a* – electric arc metallization; *b* – gas-flame spraying; *c* – hypersonic metallization; *1* – compressed air; *2* – gas mixture; *3* – wire being sprayed; *4* – nozzle; *5* – fused wire end; *6* – flow of sprayed particles; *7* – coating; *8* – substrate.

To assess advantages of coatings at friction compared to cast materials of the same compositions we carried out tribotests of the coatings sprayed from wires made of different bronzes [6]. The wear rate and friction coefficient of coated specimens were lower than those of cast specimens. The load range within which normal mechano-chemical wear runs for the coatings is by 30% wider than for the cast alloys.

The introduction of dispersed metallic particles 0.1–5.0 μm in size into greases reduces somewhat the friction coefficient of coatings sprayed from both steel wires and nonferrous alloys. The presence of the modifier in greases reduces the average temperature of surface layers by 15–20 °C and the friction coefficient by 15%. We note that the use of the modifier affects the duration of the non-stationary running-in period. The analysis of the data shows that, when applying a grease, the process stabilizes already after 15 min of operation under a pressure of 3.7 MPa at a sliding velocity of 3.82 m/s. However, the introduction of 2% of copper powder into the grease prolongs this period more than thrice (up to 50 min). Additionally, the temperature in the friction zone elevates more gradually. This can be explained by the fact that the presence of ductile particles in the friction zone retards changes in the friction surface morphology and the formation of quasi-stable microrelief. However, the modifiers in the form of particles of the above-mentioned size did not provide the efficiency of gas-thermal coatings under pressures over 25–30 MPa.

To widen the operation range of gas-thermal coatings from wire materials towards heavier pressures (up to 100 MPa) we proposed to apply lubricants modified by diamond nanoparticles.

Coatings sprayed from martensite an austenite steels were tested at reciprocal sliding under conditions of boundary lubrication. The pressure was 10–100 MPa and the average velocity of relative motion was ≈ 0.1 m/s. The counterbody was made of hardened steel U8 (HV = 7800–8000 MPa). The friction pair was lubricated with commercial oil I-20A which was modified by nano-sized diamond-graphite charge (UDDG). The concentration of UDDG in the oil was 1 wt.%. Phase and structural transformations in surface layers of the coatings were studied using X-ray analysis.

Results of the tribotests have shown that, when using the non-modified oil I-20A, the friction coefficient of the coatings in pair with steel U8 (the steel had the following chemical composition, wt.%: C – 0.80, Si – 0.27, Mn – 0.27, Cr – 0.20, Ni – 0.25, Cu – 0.25, and the rest was Fe) was 0.11–0.12 at initial stages of testing. Owing to running-in it decreased down to 0.095 for HM-coatings of austenite stainless steel 12Kh18N10T (the steel had the following chemical composition, wt.%: C – 0.12, Si – 0.80, Mn – 0.20, Cr – 18.2, Ni – 10.5, Ti – 0.60, Cu – 0.30, and the rest was Fe) after passing a sliding distance of 2 km and down to 0.090 for HFS-coatings of martensite steel 40Kh13 (the steel had the following chemical composition, wt.%: C – 0.40, Si – 0.80, Mn – 0.80, Cr – 13.0, Ni –0.60, Cu – 0.30, Ti – 0.20, and the rest was Fe) after passing a sliding distance of 1.6 km. The introduction of 1 wt.% of the nano-sized diamond-graphite modifier into the oil caused the acceleration of running-in. After passing a sliding distance of 2 km the friction coefficient was 0.080 for the coatings of steel 12Kh18N10T and 0.075 for the coatings of steel 40Kh13 (figure 3.13).

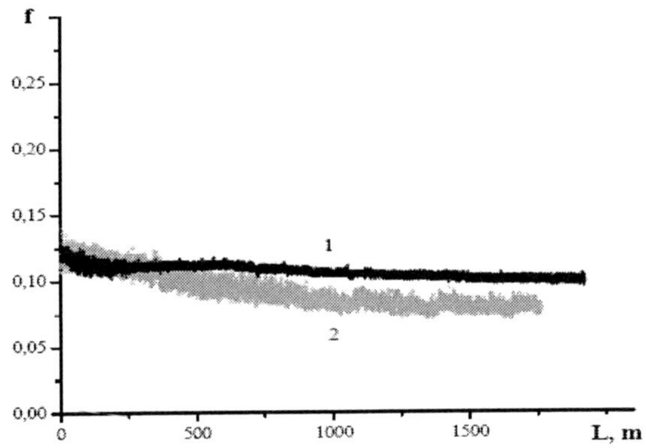

a

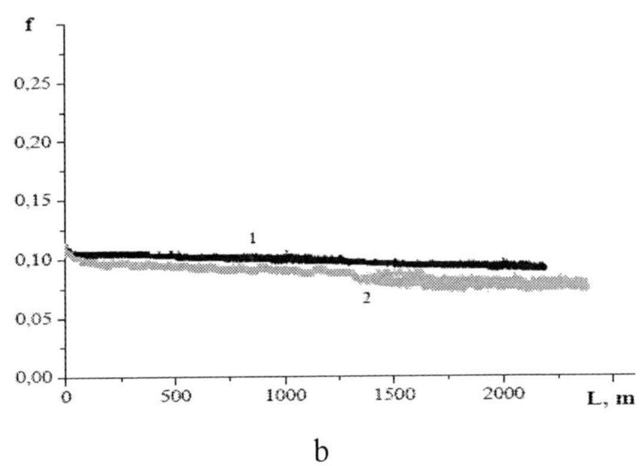

b

Figure 3.13. Dependence of friction coefficient on sliding distance for HM- (a) and HFS- coatings (b) from steel 40Kh13: *1* – oil I-20A; *2* – oil I-20A + 1% UDDG (pressure 20 MPa, counterbody – hardened steel U8).

When testing the coatings in oil I-20A under heavy contact pressures (≥ 50 MPa), the scoring of the pair occurred followed by transition to catastrophic wear (figure 3.14, a).

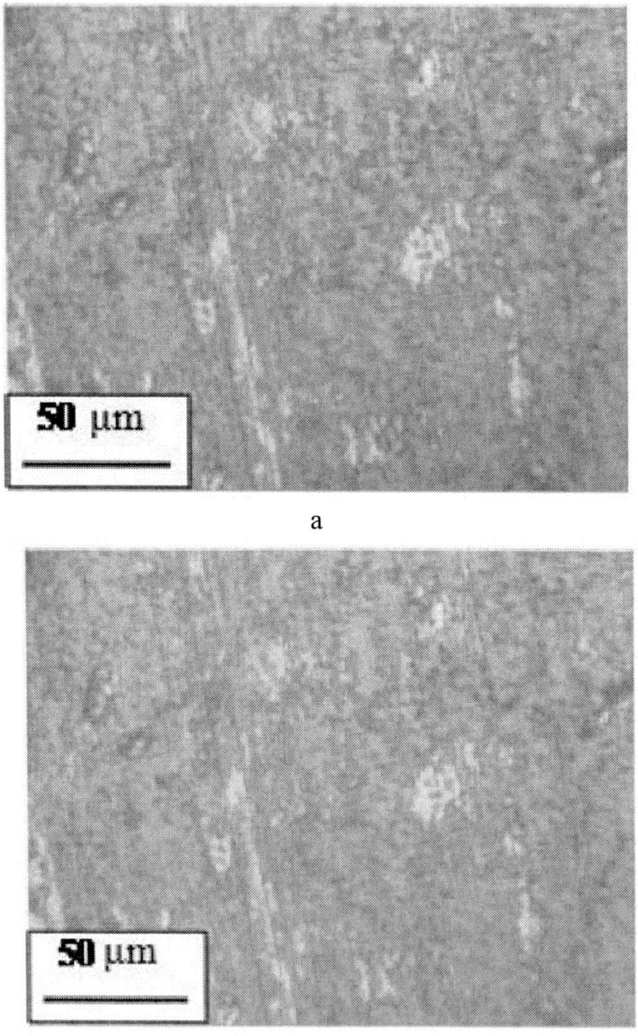

a

b

Figure 3.14. Microstructure of friction surface of coatings from steel 12Kh18N10T after lubrication with oil I-20A (*a*) and oil I-20A modified by nanodiamonds (*b*) under pressure of 100 MPa.

When testing in the modified oil I-20A, the steel 12Kh18N10T coating retained its efficiency up to a pressure of 100 MPa (figure 3.14, b). It can be assumed that running-in processes in the surface layer of the GTS-coating accelerate when using the oil containing ultradispersed diamonds. These processes are accompanied by the martensite $\gamma \rightarrow \alpha$ transformation and the appearance of interlayers with a nano-sized fragmented structure, which hampers the propagation of microcracks.

Thus, results of studies of the tribological behavior of the coatings have shown that increase in the pressure up to 50 MPa leads to the catastrophic wear of the friction pair. When using the oil modified by the additive of nano-diamonds, the running-in of the pair accelerates. The carrying capacity of the pair with the gas-thermal coating augments considerably and the pair operates in a stable manner under contact pressures up to 100 MPa.

REFERENCES

[1] Tribological Properties of Antifriction Self-Lubricating Plastics; ed. G.V. Sagalaev and N.L. Shembel; Standard Publishing: Moscow, SU, 1982; 64 p.

[2] Tseev, N.A., Kozelkin, V.V. et al. Materials for Dry Friction Units Operating in Vacuum; Mashinostroenie: Moscow, SU, 1991, 192 p.

[3] Ginzburg, B.M., Tochilnikov, D.G. et al. *Journal of Applied Chemistry*. 2006, Vol 79, N 5, 705-716.

[4] Vityaz, P.A., Zhornik, V.I. et al. *Heavy Engineering*. 2005, N10, 19-22.

[5] Vityaz, P.A., Zhornik, V.I. et al. *Journal of Friction and Wear*. 2006, Vol 27, N 1, 61-68.

[6] Belotserkovsky, M.A. Activated Flame Spraying. Monograph; Technoprint: Minsk, BY, 2004; 200 p.

[7] Hocking, M.G., Vasantasree V., and Sidky P.S. Metallic and Ceramic Coatings; Longman Group UK Limited: London, UK, 1989; 516 p.

[8] Belotserkovsky, M. Priadko, A. Hardening Technology and Coatings, 2006, N12, 17-23.

Chapter 4

Formation of Wear-Resistant Surface Structures at Tribomechanical Treatment in Lubricant Containing Hard Nano-Sized Components

4.1. Structure and Phase Transformations in Surface Layer of Material Lubricated by Grease Modified with Nanodiamonds

To study peculiarities of structure and phase transformations in surface layers of materials lubricated by the modified grease we carried out tribotests of the steel 45 – steel 45 and copper M1 – steel 45 pairs.

Originally, annealed steel 45 has the ferrite-pearlite structure with the lattice spacing of the α-phase a_α = 0.2866 nm and contains particles of orthorhombic carbide Fe_3C (figure 4.1, *a*). Diffraction lines of the matrix α-phase are narrow that proves a low density of defects in the ferrite lattice of annealed steel 45. The microhardness of electropolished steel is H_μ = 1800 MPa. Table 4.1 contains the data on the structure state and microhardness of surface layers of annealed steel 45 after tribotesting under various conditions.

Table 4.1. Microhardness H_μ and Structure Parameters of Annealed Steel 45 after Tribotesting

Material	**Test conditions**	H_μ, MPa	β_{110}, 10^{-3} rad	β_{220}, 10^{-3} rad	β_{220}/β_{110}	D, µm
Annealed steel 45	10 MPa Litol-24	2350	1,9	2.9	1.6	0.13
	10 MPa Litol-24 + 1% UDDG	2700	2.3	3.8	1.7	0.10
	20 MPa Litol-24	2500	2.2	3.2	1.5	0.12
	20 MPa Litol-24 + 1% UDDG	2950	2.5	4.0	1.6	0.09
	30 MPa Litol-24	2700	2.3	4.1	1.8	0.09
	30 MPa Litol-24 + 1% UDDG	3300	3.3	6.0	1.8	0.06

After lubrication by the grease Litol-24 the annealed steel surface becomes smooth and acquires high luster (figure 4. 2, *a*). Under high magnification the friction surface of annealed steel lubricated by the grease Litol-24 modified with UDDG shows smoothed plasticized areas (figure 4.2, *b*).

As the data presented in Table 4.1 show, after the tests the microhardness of the surface layer of annealed steel increases considerably (1.5–1.8 times) and the diffraction lines of the matrix α-phase become wider. This proves a substantial increase in the density of dislocations and vacations in steel surface layers due to their severe plastic deformation at friction.

The ratio β_{220}/β_{110} equals to 1.5–1.8 that is quite close to the ratio $sec_{220}/sec_{110} = 1.9$. This value of the ratio β_{220}/β_{110} argues for the formation of disoriented dislocation substructures in surface layers [1, 2]. The structures are characterized by the correlated arrangement of dislocations as walls consisting of equidistant dislocations, which have the same sign [8]; areas inside the walls are slightly distorted and contain almost no dislocations. This structure is typical for disoriented cells, blocks, or subgrains and the value of physical widening is inversely proportional to the size of blocks D [3].

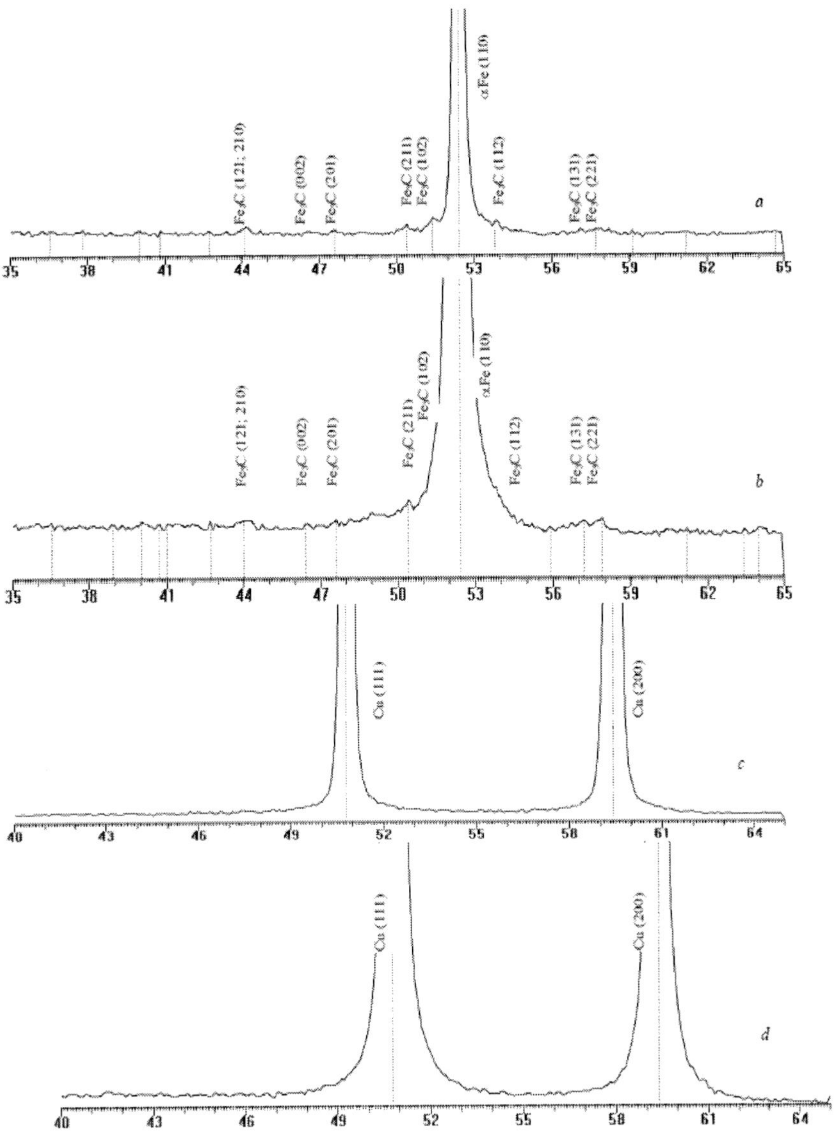

Figure 4.1. Fragments of diffraction pictures (CoK_α) from surface layers of annealed steel 45 (*a*, *b*) and copper M1 (*c*, *d*): *a*, *c* – original state (after electrochemical polishing); *b*, *d* – after lubrication with grease Litol-24 modified by UDDG under pressure of 20 MPa.

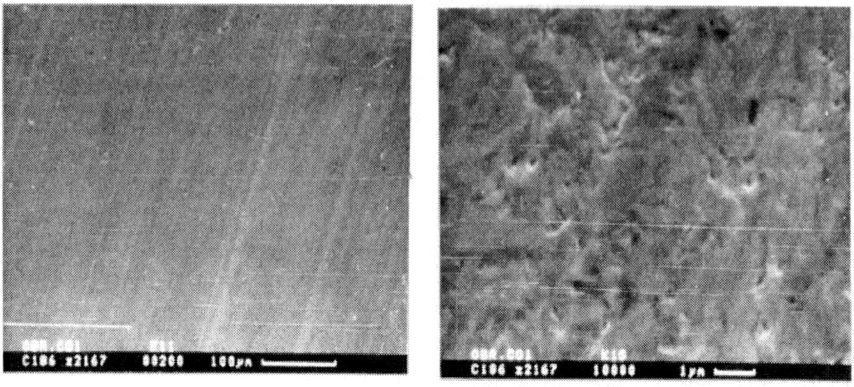

a b

Figure 4.2. Microstructure of friction surface of annealed steel 45 after lubrication with grease Litol-24 modified by UDDG under pressure of 20 MPa at various magnifications.

Figure 4.3, *a* shows a typical micrograph of the subgrain structure appeared in surface layers of annealed steel 45 lubricated by the grease modified with UDDG. The formation of the subgrain structure leads to a considerable increase in the friction surface hardness. The comparison of the data on structure parameters of steel friction surfaces lubricated by the modified and nonmodified greases shows that the presence of the UDDG additives in the grease results in a significant rise of the physical widening of diffraction lines of the matrix α-phase. This proves the formation of the dispersed subgrain structure in thin surface layers of steel 45 lubricated by the grease modified with UDDG. The structure contains essentially smaller blocks compared to the structure formed under lubrication with nonmodified grease. In fact, during the tests with the grease modified by UDDG, the nano-sized cellular substructure appears which has the subgrain size of $\leq$ 100 nm. With increasing the pressure, the value of β grows and the difference between the values registered in the tests with the modified and nonmodified greases Litol-24 becomes more considerable. The refinement of the subgrain structure appeared in friction is characterized by the block size *D* and accompanied by a proportional increase in the friction surface microhardness (Table 4.1). In particular, during the tests under a pressure of 30 MPa the surface layer microhardness grows approximately twice. The formation of the nano-sized subgrain structure in surface layers owing to its extremely high ductility results in effective frictional energy adsorption and facilitates the running-in of the pair.

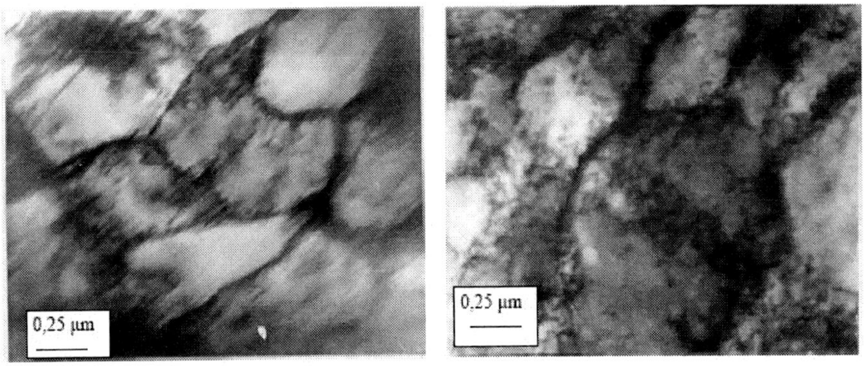

Figure 4.3. Microstructure of friction surface of annealed steel 45 (*a*) (pressure 20 MPa) and copper M1 (*b*) (pressure 10 MPa) after lubrication with grease Litol-24 modified by UDDG.

Additionally, such structures are characterized by a high resistance to the nucleation and propagation of microcracks [4, 5] that yield wear particles. From the other hand, the severe work-hardening of the nano-sized subgrain structure provides a high carrying capacity of the friction pair. It can be assumed that the effective refinement of the subgrain structure appeared in surface layers when lubricating with the grease modified by UDDG favors the improvement of the tribological properties of such friction pairs. A possibility of the direct modification of ductile materials of the pair by ultradispersed diamonds should also not be excluded.

The study of durometric properties and the structure state of copper has shown that electropolished copper possesses low microhardness (H_μ = 650 MPa) and density of lattice defects. The X-ray diffraction picture shows narrow lines of the matrix γ-phase (figure 4.1, *c*). The lattice parameter of the matrix phase is a_γ = 0.3615 nm. The mechanical grinding of copper specimens during their preparation to tribotesting leads to a considerable increase in the surface layer microhardness to H_μ = 900–1000 MPa and the number of lattice defects such as dislocations, vacations, and others. This is proven by the growth of the physical widening of diffraction lines (β_{222}CoK$_\alpha$ = $10.53 \cdot 10^{-3}$ rad). The lattice parameter is a_γ = 0.3616 nm.

Table 4.2 represents the data on the structure parameters and microhardness of friction surfaces of copper M1 specimens after tribotesting.

Table 4.2. Microhardness and Structure Parameters of Copper M1 after Tribotesting

Material	Test conditions	H_μ, MPa	β_{111}, 10^{-3} rad	β_{222}, 10^{-3} rad	β_{220}/β_{110}	D, μm
Copper M1	10 MPa Litol-24	1200	2.3	3.8	1.7	0.09
	10 MPa Litol-24 + UDDG	1500	2.9	4.6	1.6	0.08
	20 MPa Litol-24	1300	2.7	4.9	1.8	0.07
	20 MPa Litol-24 + UDDG	1600	3.1	5.2	1.6	0.06

It can be seen that owing to friction the physical widening of diffraction lines of the matrix γ-phase drops sharply (from $\beta_{222} = 10.53\cdot10^{-3}$ rad to $\beta_{222} = (4–5)\cdot10^{-3}$ rad). This transformation is caused by the appearance of the subgrain structure in surface layers. The data presented in Table 4.3 also show that copper demonstrates regularities of the formation of the surface layer structure in friction similar to those typical for annealed steel 45. In particular, with increasing the pressure the microhardness of copper grows significantly and the subgrain structure is refined. The presence of the UDDG additives in grease Litol-24 intensifies the work-hardening of surface layers of copper M1 and reduces the size of subgrains appeared (figure 4.3, *b*).

Thus, it has been found that during the tribotests of the friction pairs ductile material – hard material lubricated with the grease Litol-24 modified by UGGD, the nano-sized subgrain structure is formed in surface layers of ductile materials, i.e. annealed steel and copper. The structure is characterized by a smaller subgrain size (≤ 100 nm) and an increased microhardness of surface layers that improves the wear resistance of the friction surface [6].

4.2. Variations in Topography of Friction Surface Lubricated by Grease with Nanodiamonds

The study of the friction surface topography by optical and scanning electron microscopy has shown that after sliding tests with the commercial grease Litol-24 under a pressure of p_a = 20 MPa the surface of hardened and annealed steel 45 becomes smooth and acquires high luster (figures 4.4, *a*, *b* and 4.5, *a*, *b*). The surface demonstrates stripes resulted from the plastic deformation of thin surface layers at friction against the counterbody. The microhardness of surface layers of annealed steel 45 after friction under a pressure of p_a = 20 MPa reaches H_μ = 2500 MPa. For steel hardened in water the surface microhardness remains similar to that of original steel (H_μ = 2500 MPa).

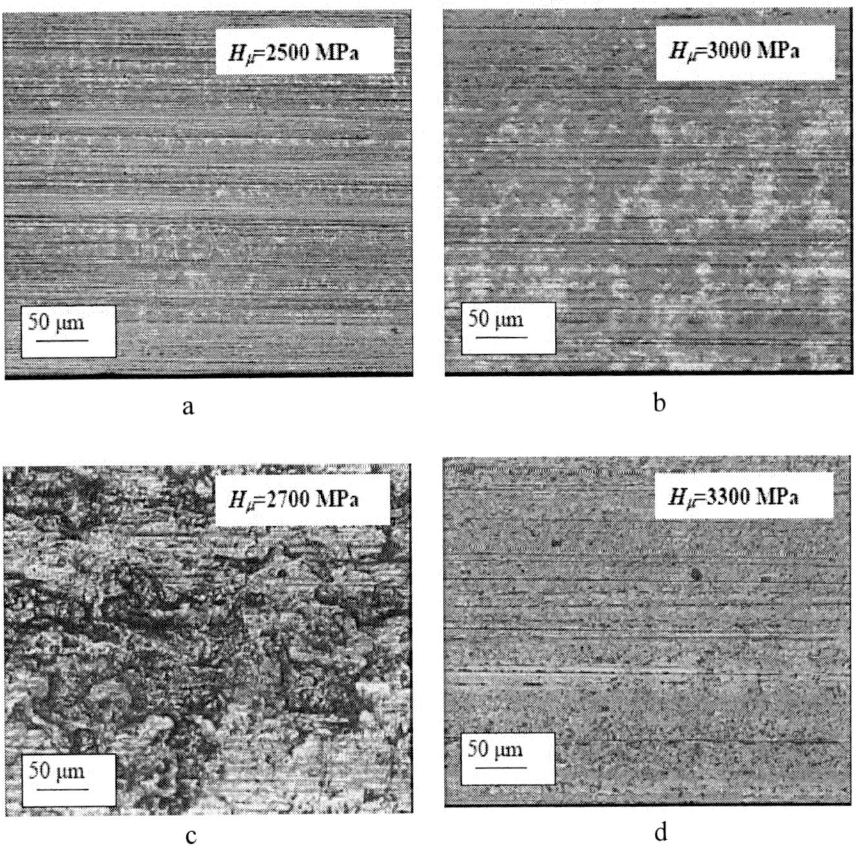

Figure 4.4. Microstructure of friction surface of annealed steel 45 with initial hardness of H_μ = 1800 MPa after tribotests: *a* – Litol-24, pressure p_a = 20 MPa; *b* – Litol-24+UDDG, pressure p_a = 20 MPa; *c* – Litol-24, pressure p_a = 30 MPa; *d* – Litol-24+UDDG, pressure p_a = 30 MPa.

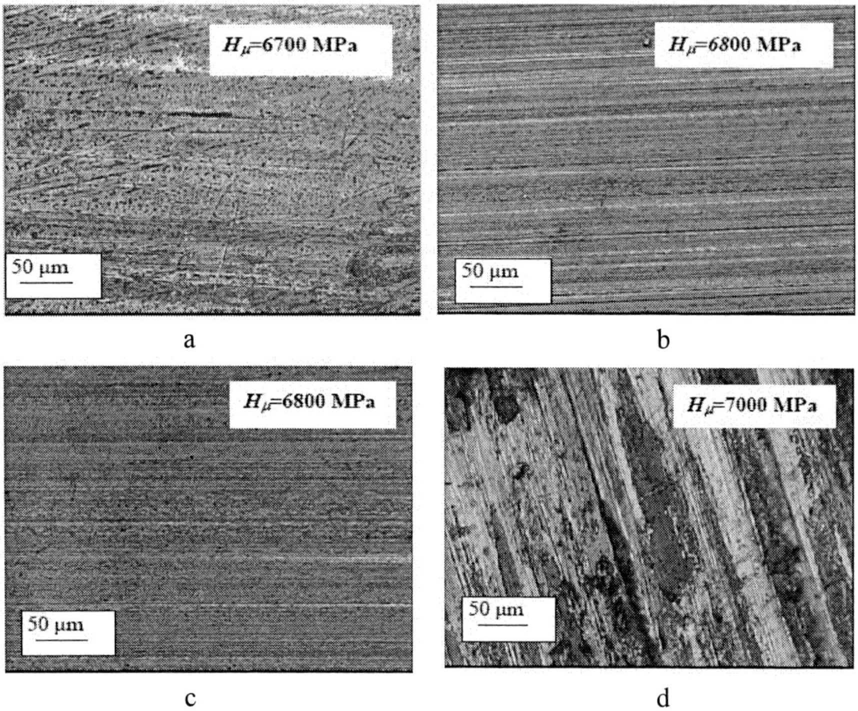

Figure 4.5. Microstructure of friction surface of hardened steel 45 with initial hardness of H_μ = 6400 MPa after tribotests: *a* – Litol-24, pressure p_a = 20 MPa; *b* – Litol-24+UDDG, pressure p_a = 20 MPa; *c* – Litol-24, pressure p_a = 30 MPa; *d* – Litol-24+UDDG, pressure p_a = 30 MPa.

The modification of grease Litol-24 by ultradispersed diamond-graphite charge UDDG reduces the friction surface roughness of specimens of hardened and annealed steel tested under a pressure of p_a = 20 MPa. After the tests under a pressure of p_a = 20 MPa the friction surface microhardness increases up to $H_\mu \approx 3000$ MPa for annealed steel 45 and up to H_μ = 6800 MPa for hardened steel.

Increase in the nominal contact pressure up to p_a = 30 MPa leads to significant changes in the friction surface topography of steel specimens lubricated with the commercial grease Litol-24 and the grease Litol-24 modified by UDDG. In particular, the lubrication of annealed steel 45 with the nonmodified grease Litol-24 under heavy contact pressures (p_a = 30 MPa) is accompanied by the plasticization of the surface layer and its severe wear due to adhesion to the counterbody material [7].

In case of hardened steel the influence of the UDDG additives is more complex. If the initial hardness is relatively low (H_μ < 4500 MPa), the use of the modified grease improves tribological properties of the pair, while at a higher initial hardness the positive effect of the UDDG additives flags (figure 4.6).

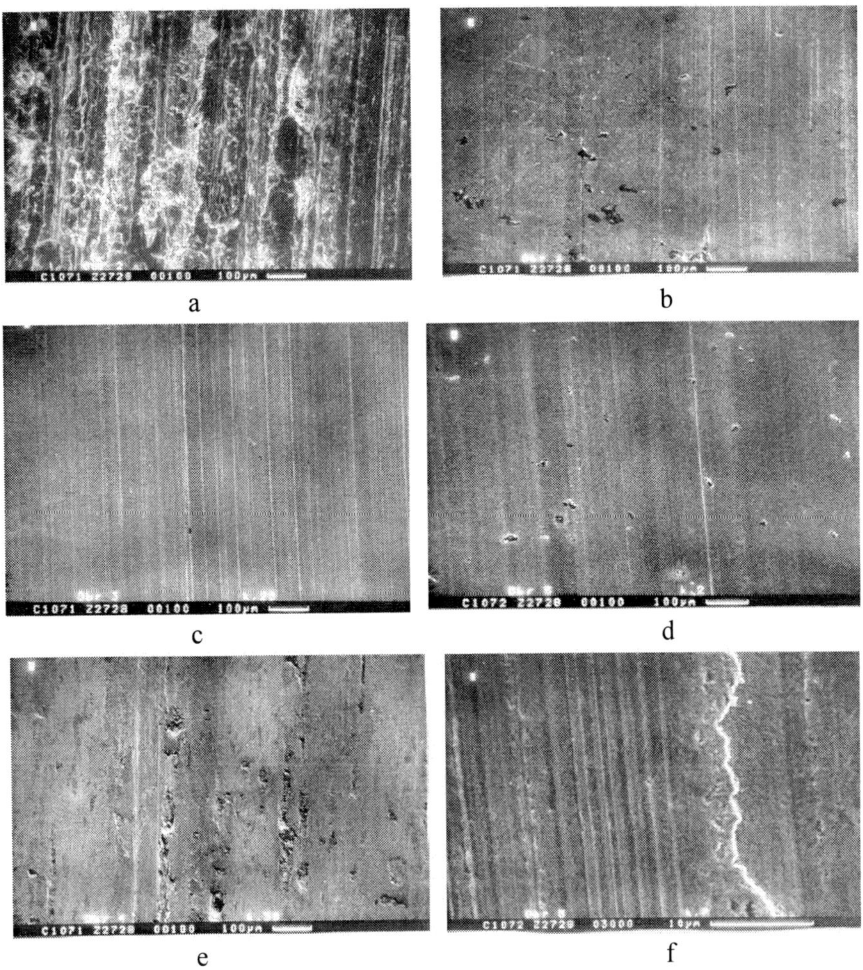

Figure 4.6. Topography of friction surface of steel 45 with various hardnesses after lubrication with commercial and modified greases under pressure of p_a = 30 MPa: *a* – *HV* = 1800 MPa, Litol-24; *b* – *HV* = 1800 MPa, Litol-24+UDDG; *c* – *HV* = 4500 MPa, Litol-24; *d* – *HV* = 4500 MPa, Litol-24+UDDG; *e, f* – *HV* = 5500 MPa, Litol-24+UDDG.

The friction surface shows the areas with a great number of microcracks and deep hollows resulted from the separation of wear particles (figures 4.4, *c* and 4.6, *a*). The surface microhardness of annealed steel 45 reaches H_μ = 2700 MPa. The introduction of ultradispersed diamonds into grease Litol-24 leads to an increase in the friction surface microhardness of annealed steel 45 up to H_μ = 3300 MPa. The friction surface becomes smoother (figure 4.4, *d*). Specimens of hardened steel with an initial hardness of H_μ > 4500 MPa lubricated with commercial grease Litol-24 show the striated smooth relief (figure 4.5, *c*). The layer microhardness is H_μ = 6800 MPa. The modification of grease Litol-24 intensifies the wear of hardened steel 45. It is seen most noticeably if the initial hardness of steel is high (H_μ = 6400 MPa); friction is accompanied by the appearance of severe wear areas with the rough surface relief (figure 4.5, *d*) [8, 9].

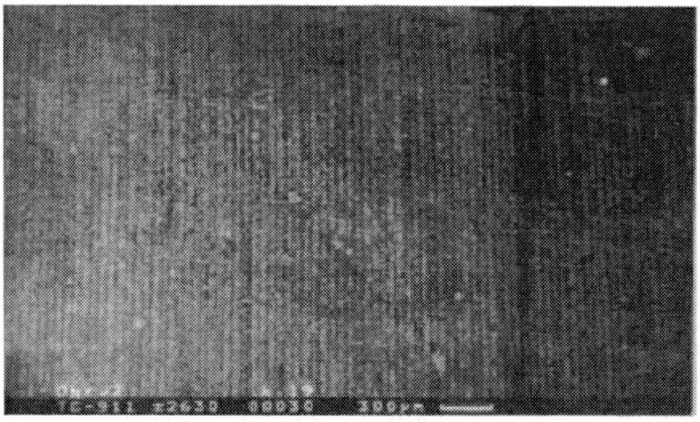

a

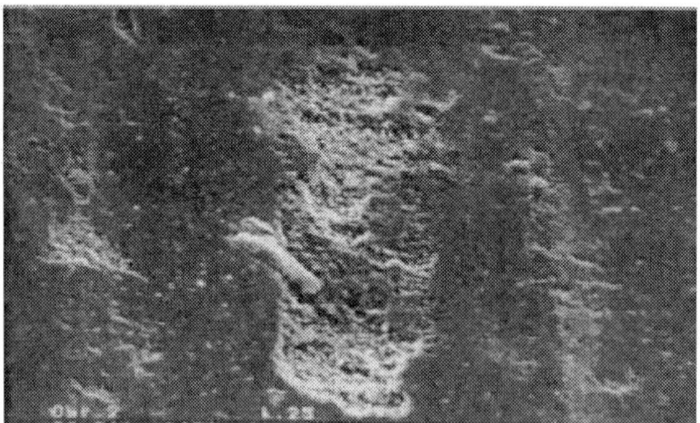

b

c

d

Figure 4.7. Microstructure of friction surface of steel ShKh-15 with hardness of HRC 35–40 after lubrication with modified (*a*, *b*) and commercial (*c*, *d*) grease Litol-24.

The examination of the friction surfaces by optical and electron microscopy carried out after rolling tests has shown that the friction surface of steel ShKh-15 with an initial hardness of 35–40 HRC lubricated with the modified grease is smoother. It demonstrates single rounded pits about 10×40 μm in size that contains no cracks (figure 4.7, *a*, *b*).

When using Litol-24 without nano-sized diamond-containing additives, pits appear over relatively vast areas and are covered with a net of chisel cracks that proves the severe failure of the friction surface (figure 4.7, *d*). In this case the positive role of the nano-diamond additives with increasing the friction surface microhardness becomes less important [10].

4.3. Model of Friction Failire of Surface Lubricated by Grease with Hard Nano-Sized Components

It has been shown in Sections 4.1 and 4.2 that under light contact pressures the modification of greases by UDDG additives accelerates substantially the running-in of friction pairs. Under heavy contact pressures (p_a = 20–30 MPa) the efficiency of the modification of grease by UDDG additives depends much on the initial hardness of the contacting materials. In case of steel 45 with a high hardness the UDDG additives intensify processes of microfailure at friction against the counterbody and accelerate transition to the severe wear stage.

The mechanism of friction failure in the presence of nano-sized diamond particles can be considered from the viewpoint of linear fracture mechanics. It is known that a friction surface experiences alternating loading. A zone of compressive stresses σ_x is formed at the leading edge of the contact area, while a zone of tensile stresses appears at the trailing edge [11]:

$$\sigma_x = 2f p_0, \qquad (4.1)$$

where f is the friction coefficient and p_0 is the real pressure.

As the real contact area does not exceed ≈ 0.01–0.1% of the nominal contact area [12], alternating stresses in surface layers reach quite great values ($\sigma_x \geq \sigma_{0.2}$) that, as a rule, top much the yield point of the contacting materials. Since diamond particles are embedded in the mated surfaces during friction, a hard diamond particle in the friction surface can be considered as a round cut in a stretched plate (figure 4.8).

When such round cut with a diamond particle enters the zone of action of alternating stresses, stress concentration occurs at the cut edge owing to a small size of the particle ($r \approx 10$ nm); it reaches $\alpha = 3$ [13]. Such high stress concentration at the cut mouth induces severe local plastic deformation in this zone during friction. Subsequently, as the number of cycles tension–compression affecting the site with the embedded particle increases, a microcrack starts propagating from the round cut edge in the direction normal to the tensile stresses. Since the microcrack width is incomparable to the size of a contact spot, the plane deformed state occurs at the microcrack mouth. At the stage of the stable (undercritical) growth of cracks, when their length increases in the process of friction, the intensity K_I of the stresses acting at crack mouths rises [14]. According to ideas of linear fracture mechanics, when a crack reaches a length l_c typical for the given material, the stress intensity K_I attains its critical value $K_I = K_{IC}$ that is limiting for the material; at this value of the stress intensity the crack losses its stability [14] and starts propagating fast into deeper layers. Crack propagation yields a wear particle, i.e. it causes friction failure. It can be assumed that catastrophic wear which occurs at late testing stages is due to the fact that surface microcracks resulted from friction reach their critical size. Thus, the critical crack size is an important parameter that characterizes the capability of a material to resist friction failure in the presence of abrasive particles.

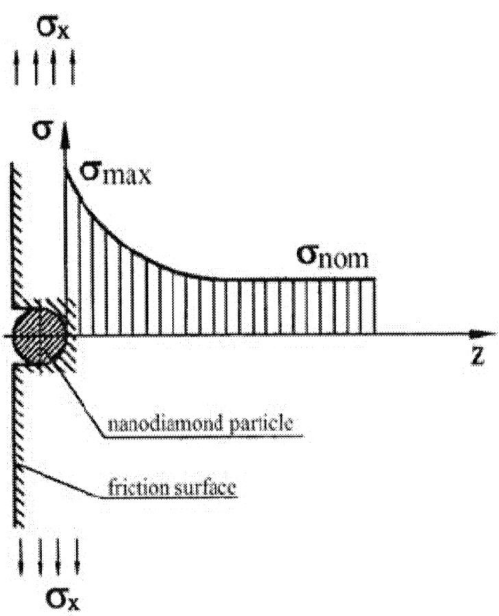

Figure 4.8. Diagram of stress distribution in vicinity of diamond nanoparticle embedded into friction surface.

The critical crack size l_c is determined in accordance with the Griffith's model modified by Orovan [14]:

$$l_c = 2\gamma_{ef}E/\pi\sigma^2, \tag{4.2}$$

where E is the elastic modulus of the material; σ is the applied stress; γ_{ef} is the effective surface energy.

The critical stress of crack motion σ_c is

$$\sigma_c = (2\gamma_{ef}E/\pi l)^{1/2}. \quad (4.3)$$

In accordance with [14, 15] $K_I = \sigma(\pi l)^{1/2}$. By substituting this expression in the condition $K_I = K_{IC}$, we obtain the following simple expression for the assessment of the initiation stress of a crack with the given length $l = l_c$:

$$\sigma_c = K_{IC}/(\pi l_c)^{1/2}. \quad (4.4)$$

Since l_c depends on the fracture toughness K_{IC}, increase in the latter characteristic causes a significant rise of l_c and the stage of the undercritical retarded crack growth becomes longer. As a result, one may expect that the failure process in friction will be hampered. For this reason it was of interest to plot critical fracture diagrams representing the dependences of the critical stress σ_c on the critical crack length l_c for steel 45 heat treated under various conditions and to compare these data with the tribotest results. We used formula (4.4) of linear fracture mechanics describing relation between the critical stress σ_c and the critical crack length l_c.

Since no experimental values of K_{IC} are available for the studied conditions of the heat treatment of steel 45, we assessed this characteristic using the following expression from [16]:

$$K_{IC} = \sqrt{\rho \tau_s E(1-\nu^2)^{-} \ln(1-\psi)^{-}}, \quad (4.5)$$

where ρ is the characteristic parameter of the structure taken, according to [21], equal to the size of a primary austenite grain; τ_s is the yield point at shear; E is the elastic modulus; ν is the Poisson's ratio; ψ is the relative contraction at rupture. The value of τ_s was found from the expression $\tau_s = \frac{1}{\sqrt{3}}\sigma_y$, where σ_y is the yield point of steel.

The authors of [16] derived the above expression based on the Leonov–Panasyuk model developed to determine the cracking resistance of ductile materials, which operate in the range of elastoplastic deformations [17]. They derived analytically the formula for the assessment of the fracture toughness K_{IC} using the basic mechanical characteristics of the material with account for the structure parameter. Estimates of K_{IC} carried out with the use of formula (4.5) for annealed and hardened steel 45 (Table 4.3) correlate quite well with the available data [16, 18].

Table 4.3. Initial Data and Results of Calculation of Critical Stress Intensity Factor K_{IC} for Annealed and Hardened Steel 45

Conditions of heat treatment of steel 45	Yield point at compression $\sigma_{0,2}$, MPa	Yield point at shear τ_s, MPa	Relative contraction at rupture ψ, %	Characteristic structure parameter ρ, mm	Critical stress intensity factor K_{IC}, $MPa{\cdot}m^{1/2}$
Annealing	450	260	65	0.03	43
Hardening in water	1200	690	4	0.03	14

Figure 4.9 shows the critical fracture diagrams plotted for annealed and hardened steel 45 based on the obtained values of K_{IC} (Table 4.3). It is seen from these data that for hardened steel 45 the critical crack size ($l_c \approx 40$ μm) at the yield stress is approximately 10_2 times less than for annealed steel 45. This means that the period of the retarded (undercritical) crack growth for hardened steel is much shorter than for annealed steel.

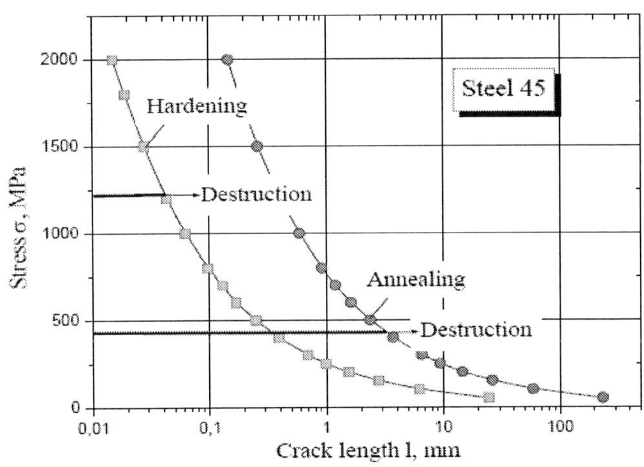

Figure 4.9. Critical fracture diagrams for annealed and hardened steel 45.

This result agrees completely with the data presented in Section 4.2 and allows us to propose the mechanism of the friction failure of materials with different hardnesses lubricated with grease modified by UDDG. It involves the embedding of diamond nanoparticles into surface layers of hardened steel, which may initiate the nucleation of fatigue microcracks in them. The cracks reach its critical size l_c quite fast and accelerate the separation of wear particles leading to the catastrophic wear of high-strength friction members (figure 4.10, *a*).

For annealed steel, in contrast, the embedding of ultradispersed diamonds into the surface hardens surface layers and accelerates processes of multiple slip in them, which result to the formation of the submicrocrystalline structure (figure 4.10, *b*) characterized by a high resistance to the propagation of microcracks [4]. The accelerated formation of hardened surface layers and the running-in of the friction pair prevent the adhesion of the ductile material to the counterbody under heavy contact pressures. Microcracks appeared in the surface layer of the ductile material under the effect of cyclic tension–compression loading or resulted from the embedding of diamond particles into the friction surface grow slowly owing to a high l_c typical for such materials and stress relaxation due to plastic deformation in microcrack mouths. For this reason the frictional energy is absorbed effectively and the pair whose members are made of ductile materials operates steadily.

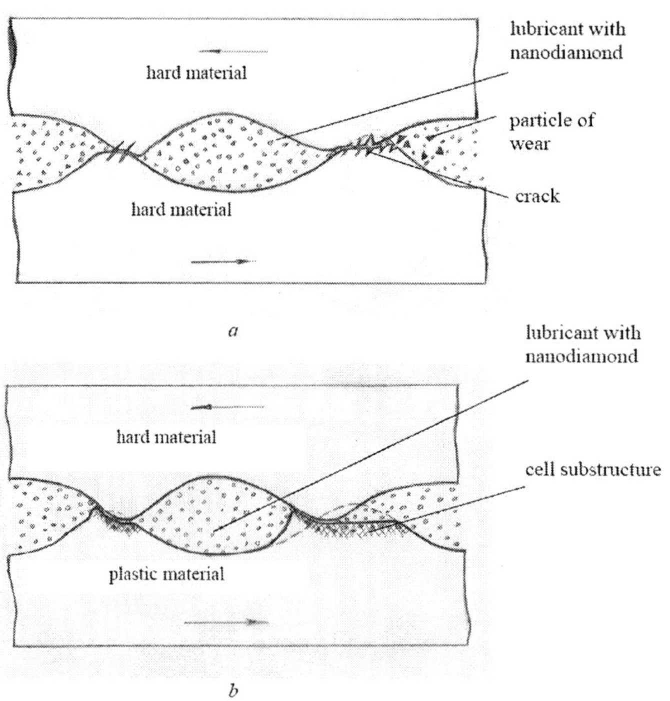

Figure 4.10. Schematic of interaction and wear particle formation during friction of materials with different hardnesses in presence of grease modified by UDDG additives.

Thus, we conclude that the presence of superhard ultradispersed diamond particles in grease may initiate the nucleation and accelerated propagation of microcracks in high-strength friction members that possess a poor cracking resistance. When a friction member is made of a more ductile material, the presence of ultradispersed diamonds in grease leads to the formation of the nano-sized cellular structure in the material surface layers. The structure has a higher resistance to the nucleation and propagation of microcracks and absorbs effectively the frictional energy that improves the wear resistance of the pair.

4.4. Structure and Phase Transformations in Surface Layer of Gas-Thermal Coatings Lubricated with Lubricant Modified by Nanodiamonds

The diffusion-free $\gamma \to \alpha$ transformation that occurs in carbon steels during accelerated cooling from the hardening temperatures leads to the appearance of the phase having the tetragonal lattice and called martensite [19]. The formation of martensite in steels increases considerably their hardness, wear resistance, and resistance to plastic deformations. In addition to carbon steels, martensite structure may appear in carbon-free steels and alloys during the polymorphic transformation that runs according to the diffusion-free mechanism [19, 20]. In particular, martensite is a typical structure, which is formed at the low-temperature polymorphic transformation in pure metals such as Fe, Co, Ti, Zr, etc., in solid solutions on their base, and in intermetallides such as CuZn, Cu_3Al, NiTi, V_3Si, and others. The appeared martensite-like structure has an improved strength and ductility. During steel cooling the martensite transformation runs within the temperature range M_b–M_e; it starts at the temperature M_b (the temperature of the beginning of the diffusion-free $\gamma \to \alpha$ transformation) and ceases at the temperature M_e (the temperature of the end of the martensite transformation). With increasing the content of carbon and alloying elements (Mn, Cr, Ni, V, and others) in steels the temperature range of the martensite transformation shifts towards lower temperatures and at the room temperature hardened steels contain a great amount of the metastable high-temperature γ-phase or retained austenite. The completion of the martensite transformation requires either the cooling of an alloy to cryogen temperatures (cold treatment) or the additional effect of magnetic field or plastic deformation [21]. It has been shown in [22] that the plastic deformation of austenite in a nickel-iron alloy favors the elevation of the temperature M_b and martensite formation. It is generally recognized now that deformation martensite is formed when the deformation temperature is below a critical temperature M_d, which, as a rule, exceeds the temperature M_b. In general case the temperature M_d for steels and alloys correlates with the temperature M_b and is governed by the extent of the alloying of the material with elements that increase the stability of austenite. With increasing the concentration of these elements the temperature M_d decreases. In high-alloy steels the martensite transformation does not occur if deformation evolves at temperatures above M_d [23]. Martensite α- and ε-phases may appear at friction of steels and alloys containing metastable austenite. The new martensite phases resulted from severe plastic deformation at friction have a high strength and wear resistance [24] that improves substantially tribological characteristics of friction pairs. Therefore, the formation of metastable austenite structures in gas-thermal coatings is a promising way of increasing their wear resistance.

The structure of the coatings deposited by the spraying of wire materials is similar to that of powder gas-thermal coatings. At powder spraying single particles may not melt and are heated only to a pre-melting temperature, while at wire spraying the coating is formed only from melted particles (otherwise drops are not separated from a wire). It causes greater plastic deformation of particles and lower porosity of the coating compared to gas-flame powder spraying. Coating deposition is accompanied by intensive physical-chemical interaction between melted particles and burning gases of the jet as well as by the interaction of particles of the sprayed material with each other when they are arranged on the part surface. The processes of the accelerated crystallization, deformation, and tempering of deposited particles run in the coating; oxides are formed and some alloying elements burn up. The structure state and properties of coatings depend on the combination of spraying parameters in a complex manner.

Among the basic requirements for surface layers of friction members are their ductility at the initial stage of friction accelerating running-in, a high hardness and wear resistance, and good adhesion to lubricants. These requirements can be fulfilled for steels if only the two-phase structure appears in them, which contains metastable austenite with a hardness of 200–300 HV. During running-in, metastable austenite transforms to wear-resistant and hard deformation martensite (700–800 HV) owing to the $\gamma \rightarrow \alpha$ deformation transformation. The hardness of the run-in surface reaches the value that can not be achieved by common methods of coating treatment.

The obtained results have allowed us to suppose that there are spraying conditions and steels, for both the GFS and HM methods, which provide the formation of a great amount of metastable austenite in the coatings capable of experiencing the $\gamma \rightarrow \alpha$ deformation transformation during friction. As it was shown above, the formation of metastable austenite in common cast steels is achieved by special alloying as well as complex heat and thermomechanical treatment, which are often not profitable.

To provide the formation of metastable austenite with a low temperature of the occurrence of the $\gamma \rightarrow \alpha$ deformation transformation (the temperature M_d) corresponding to working temperatures of sliding friction units (300–350 K), certain conditions of coating formation should be fulfilled. They are the overheating of the wire material above the melting point, material spraying, certain cooling rate and degree of oxidation of the sprayed material. These conditions change the concentration of alloying elements in the coating.

A relation between the initiation temperature of the martensite transformation of the wire material M_b and the amount of metastable austenite in the coating has been experimentally found (Table 4.4).

Table 4.4. Concentration of Metastable Austenite in Coatings Sprayed from Steels of Various Grades

Number of group of steels	Classes of steels	Temperature M_b, K	Temperature of heating during spraying, K	Concentration of austenite in coating, vol.%
1	Alloyed structural steels	550–700	1700–2000 2100–2500 > 2600	17–25 7–15 < 6
2	Toll steels	420–540	1700–2100 2200–2500 > 2500	15–25 8–12 < 6
3	Corrosion-resistant and heat-resistant steels	70–110	1700–2000 2000–2500 > 2500	95–98 90–95 90–95

For alloyed structural steels and corrosion-resistant steels of martensite class the temperature M_b is within the range 550–700 K (group of steels no. 1). It was found during the spraying of wires made of these steels that the volume concentration of metastable austenite reaches 24% if the temperature of a wire does not exceed 2000 K.

For tool (die) steels and structural spring steels the temperature M_b is within the range 420–540 K (group of steels no. 2). When spraying these steels, metastable austenite may appear in the concentration of 15–25 vol.% if the wire is heated to a temperature no more than 2100 K.

For corrosion-resistant and heat-resistant steels as well as Hadfield steel the temperature M_b is 70–110 K (group of steels no. 3). Owing to the low initiation temperature of the martensite transformation the austenite structure of these steels possesses high stability; that's why they are called steels of austenite class. The phase hardening of surface layers of common steels of austenite class due to the martensite transformation requires either deep overcooling or high degrees of deformation, which can not be achieved during lubricated friction (boundary lubrication). It has been found that the thermal spraying of wires made of these steels at a temperature above 2500 K may increase the temperatures M_b and M_d and cause the destabilization of the austenite phase in coatings.

Reduced stability of austenite in coatings sprayed from the steels that belong to the third group at temperatures above 2500 K is explained in the following manner. The content of manganese and chrome in steel exerts the strongest influence on the position of the temperature range within which the martensite transformation evolves. For example, decrease in the manganese content from 5 to 1% leads to the elevation of the temperature M_b from 270 to 470 K. Therefore, one of the possible ways to elevate the temperature M_b is to reduce the content of chrome or manganese in the steel austenite phase by oxidizing it during spraying. Changes in the coating composition due to intensive oxidation at temperatures above 2500 K have allowed us to reduce the stability of the austenite structure and elevate the temperature M_d up to the room temperature.

The retaining of a great amount of metastable austenite when spraying wires made of the steels belonging to the first two groups can be explained as follows. A high crystallization rate of steel particles during coating spraying and a low cooling rate of the coating within the temperature range of the martensite and bainite transformations at coating cooling provide the temperature stabilization of austenite. The temperature stabilization of austenite becomes more effective if the substrate is preheated. In particular, increase in the content of metastable austenite was found when the part surface was heated to 480 K. Further part heating reduced the adhesion of the coating. Part heating before the spraying of alloyed wire steels with a high temperature M_b makes the effect of the stabilization of austenite stronger and favors the growth of the volume of the metastable austenite phase in the coatings.

Figure 4.11 shows fragments of X-ray diffraction pictures (CoK_α) from surface layers of HM-coatings of steel 12Kh18N10T in the original state (after mechanical grinding with abrasive cloth) and after tribotests in oil I-20A of the standard composition and the same oil modified by nanodiamonds.

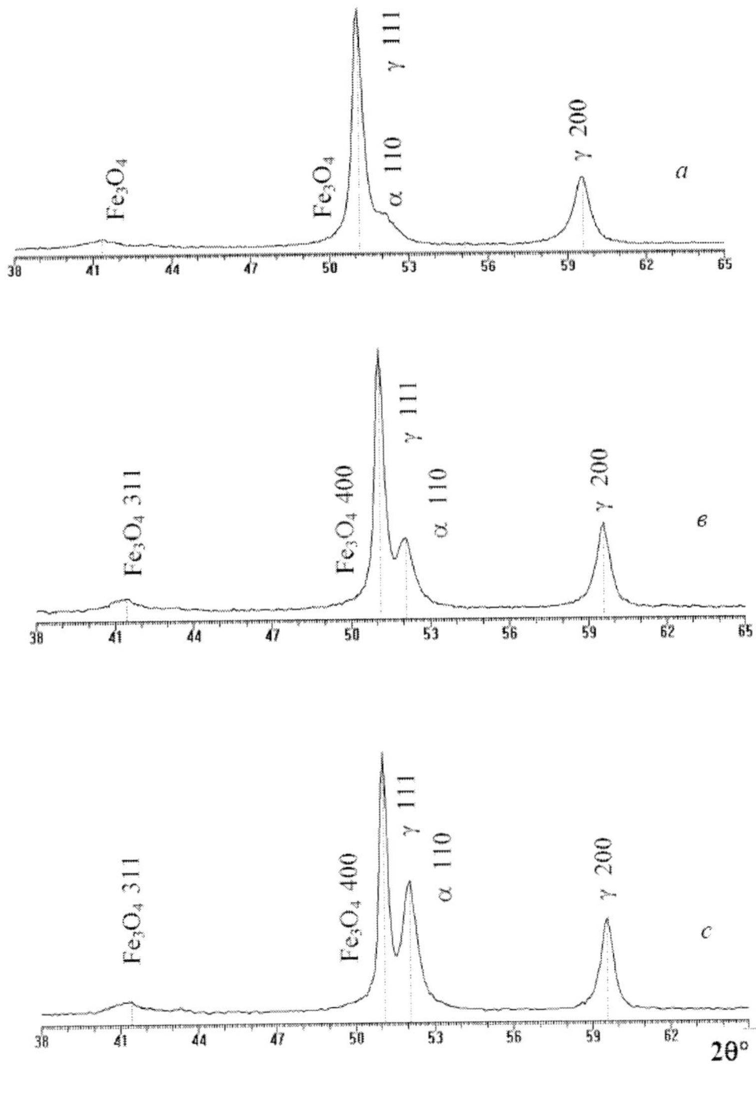

Figure 4.11. Fragments of diffraction pictures (CoK_α) from surface layers of HM-coatings of steel 12Kh18N10T in the original state (*a*) and after tribotests in oil I-20A, pressure p_a = 10 MPa (*b*) and in oil I-20A+UDDG, pressure p_a = 10 MPa (*c*).

The analysis of the X-ray diffraction pictures shows that the $\gamma \to \alpha$ phase transformation of metastable austenite to deformation martensite evolves in coating surface layers during friction. The use of the modified oil accelerates the growth of the content of the α-phase (up to 35–40 vol.%). To obtain similar amount of the α-phase when using the nonmodified oil it was necessary to increase the pressure up to 20 MPa.

Similar results were obtained during tribotests of gas-flame coatings sprayed from wires of steel 40Kh13 belonging to the class of martensite steels (figure 4.12). The introduction of nanodiamonds into the oil shortened the duration of the formation of the α-phase almost twice under lighter pressures.

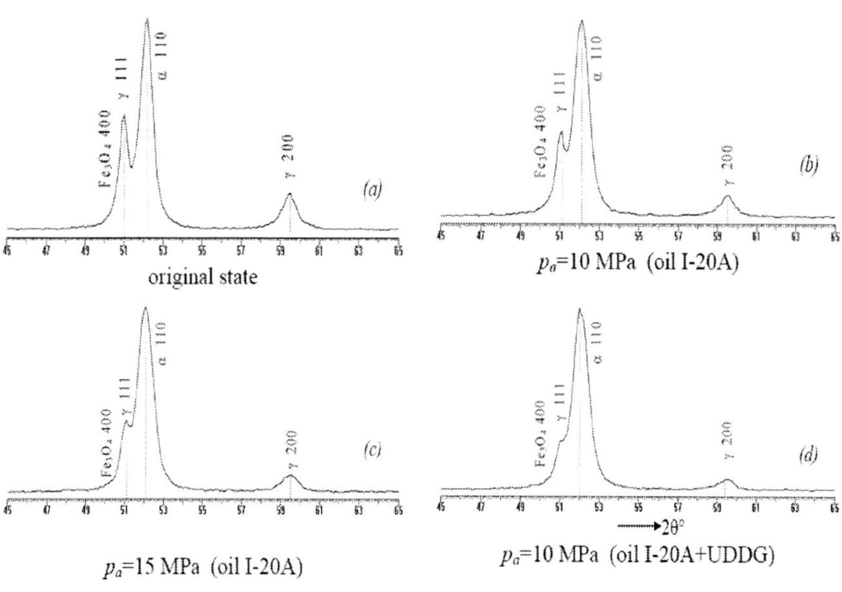

Figure 4.12. Fragments of diffraction pictures (CoK_α) from surface layers of GFS-coatings of steel 40Kh13 after lubrication with oil I-20A (*a*, *b*, *c*) and oil I-20A modified by nanodiamonds (*d*).

Subsequent studies have shown that HM-coatings of steel 12Kh18N10T lubricated with the oil modified by nanodiamonds retain their efficiency under nominal pressures of up to 100 MPa (Table 4.5). It can be assumed that, like monolytic materials, the nano-sized fragmented substructure appears in the surface layer of the coatings lubricated with the modified oil in addition to the acceleration of running-in accompanied by the $\gamma \to \alpha$ martensite transformation. The substructure provides a high resistance to the nucleation and propagation of microcracks in friction.

Table 4.5. Microhardness H_μ and Content of Retained Austenite in Surface Layers of HM- Coatings of Steel 12Kh18N10T after Friction under Various Conditions

Friction conditions	Microhardness H_μ, MPa	Content of Retained Austenite, vol.%
Original state	3300	88
p = 10 MPa, oil I-20A	5000	80
p = 10 MPa, oil I-20A + 1% UDDG	5300	75
p = 50 MPa; oil I-20A + 1% UDDG	5300	60
p = 100 MPa; oil I-20A + 1% UDDG	5500	55

REFERENCES

[1] Taylor, A. X-Ray Metallography; Metallurgiya: Moscow, SU, 1965; 410 p.

[2] Koneva, N.A., Kozlov, E.V. et al. *Physics of Metals. 1991*, Vol 12, N 1, 49-58.

[3] Krivoglaz, M.A. X-Ray and Neutron Diffraction in Non-Ideal Crystals; *Naukova Dumka:* Kiev, SU, 1983, 408 p.

[4] Armstrong, R.V. Strength of Metals with Ultrafine Grains; Ultrafine Grains in Metals; ed. by G.V. Kurdyumov; *Metallurgiya:* Moscow, SU, 1973; pp 11-40.

[5] Rybin, V.V. Severe Plastic Deformations and Fracture of Metals; *Metallurgiya:* Moscow, SU, 1986, 224 p.

[6] Vityaz, P.A., Zhornik V.I. et al. *Journal of Friction and Wear.* 2006, Vol 27, N 2, 196-200.

[7] Vityaz, P.A., Zhornik V.I. et al. Transactions of National Academy of Sciences of Belarus. *Physical-Technical Series.* 2008, N 4, 45-49.

[8] Kamko, A.I., Zhornik V.I. et al. Trends in Application of Modified Greases to Improve Serviceability of Friction Units of Agricultural Machinery; Agricultural Machinery for Harvesting of Cereals, Forage, Roots, and Tubers. State-of-the-Art, Trends, and Directions of Development; RCUE "Parent Design Bureau of Harvesting Machinery": Gomel, BY, 2007; pp 97-102.

[9] Kamko, A.I., Belotserkovskii M.A. et al. Assessment of Tribological Properties of Coatings Lubricated with Modified Oil; Ecological Safety and Energy Saving in Transport; ed. by V.M. Ovchinnikov; Belarus State University of Transport: Gomel, BY, 2006; pp 156-159.
[10] Shipitsa, N.A., Zhornik V.I. et al. Effect of Ultradispersed Diamonds on Properties of Greases; Polymer Composites and Tribology; Metal-Polymer Research Institute of National Academy of Sciences of Belarus: Gomel, BY, 2005; pp 249-250.
[11] Johnson, K.L. Contact Mechanics; Cambridge University Press: Cambridge, UK, 1987
[12] Friction, Wear, and Lubrication. Vol. 1; ed. by I.V. Kragelskii; *Mashinostroenie:* Moscow, SU, 1978
[13] Aleksandrov, A.V., Potapov, V.D. et al. Strength of Materials; *Vysshaya Shkola:* Moskow, RU, 2004
[14] Cherepanov, G.P. Brittle Fracture Mechanics; *Nauka:* Moscow, SU, 1974
[15] Pluvinazh, G. Mechanics of Elastoplastic Fracture; *Mir:* Moscow, RU, 1993
[16] Panasyuk, V.V., Andreikiv, A.E. et al. *Physical-Chemical Mechanics of Materials.* 1977, Vol 13, N 2, 120–122
[17] Panasyuk, V.V. Limiting Equilibrium of Brittle Solids with Cracks; *Naukova Dumka:* Kiev; SU, 1968
[18] Kogut, N.S. Cracking Resistance of Structural Materials; *Vischa Shkola:* Lvov, SU, 1986
[19] Gulyaev A.P. Metal Science; *Metallurgiya:* Moscow, SU, 1977, 648 p.
[20] Physical Metallurgy; ed. by P. Kan, *Nauka:* Moscow, SU, 1968, 490 p.
[21] Sadovskii, V.D., Fokina, E.A. Retained Austenite in Hardened Steel; *Nauka:* Moscow, SU, 1986.
[22] Scheil, E. Ztschr. *Anorg. Und allg Chem.,* 1932, Bd. 207, 21-40.
[23] Sagaradze, V.V., Uvarov, A.I. Hardening of Austenite Steels: *Nauka:* Moscow, SU, 1989.
[24] Chernykh, S.G., Khadyev, M.S. *Physics of Metals and Metal Science.* 1980, Vol 55, N 1, 113-120.

Chapter 5

FIELDS OF APPLICATION OF METHOD OF TRIBOMECHANICAL TREATMENT IN LUBRICANT WITH NANODIAMONDS

5.1. PROLONGING LIFE OF FRICTION UNITS BY RUNNING-IN IN LUBRICANT WITH NANODIAMONDS

One of the effective ways of applying nano-sized diamond-containing additives are the development of antifriction and antiwear lubricants that is explained by their compatibility with various synthetic and mineral oils, efficiency at very low concentrations in the base oil, and ecological safety of the carbon additive compared to metallic or fluorocarbon particles and compounds.

First works on the development and application of ultradispersed carbon diamond-containing materials, including their using in lubricants, were carried out at the Scientific and Production Amalgamation "Almaz" (Biisk, Russia). The specialists developed the technology of the fabrication of the diamond-containing additive to oils and performed wide studies and testing of the new additive under various friction conditions [1]. Mineral or synthetic oils and cooling lubricants were used as the base of the lubricants with UDDG. The base may have a complex composition and contain the necessary functional additives that provide oxidation and foaming resistance, detergent, anticorrosion, and other properties.

The application of lubricants with the ultradispersed carbon diamond-containing additive improves considerably conditions of the running-in of friction units owing to the prevention of scoring and seizure, accelerates the attainment of nominal operating conditions of mechanisms, reduces wear and prolongs the life of friction pairs. The modification of lubricants by the UDDG charge improves the serviceability of precision metal-working machinery, reduces fuel and lubricant consumption, increases the power of internal combustion engines, facilitates their start-up and reduces their noise level [2].

The production technology of oil with the UDDG modifier involves the dehydration of cluster carbon by vacuum drying, the premixing of carbon with the base oil, the filtration of the mixture through mesh filters and magnetic separators and final dispersion using, for example, disintegrators, acoustic (ultrasonic) mixers, or mills. A preferable method of producing the modified oil is the preliminary fabrication of the concentrate with 2–8 wt.% of cluster carbon according to the above procedure. Subsequently, the concentrate is diluted by the base oil to reduce the concentration of cluster carbon down to 0.01–1.0 wt.% [2].

The fabrication of greases requires the disintegration of UDDG conglomerates before thickening acids are introduced into the dispersion fluid (oil). The disintegration of conglomerates is achieved by the treatment of the UDDG charge in a reactor-mixer at the initial stage of grease fabrication; it involves the treatment of the charge by an alkali water solution followed by its heating and disintegration in oil together with sols of the disperse phase [3].

The preferable content of the nonabrasive diamond phase in cluster carbon is 30–50 wt.% that corresponds to the standard content of diamond particles in the UDDG charge. The content of the ultradispersed carbon diamond-containing charge in a lubricant depends on the pattern and conditions of operation of a device (a machine, mechanism, or unit) in which the lubricant is applied. For example, in case of heavy-duty units having mainly spacious sliding surfaces it amounts to 0.3–1.0 wt.%. For moderately loaded mechanisms the concentration of the ultradispersed carbon diamond-containing charge in the lubricant is 0.01–0.3 wt.%.

In case of lightly loaded mechanisms the concentration of the ultradispersed carbon diamond-containing charge in the lubricant is 0.01–0.15 wt.%. If the concentration of carbon clusters is below 0.01 wt.%, no considerable changes in the performance of the lubricant occur [1].

Table 5.1 presents characteristics of the running-in of new carburetor engines when using industrial oil with a viscosity of 50 cSt modified by 0.1 wt.% of cluster carbon containing 40 wt.% of nonabrasive diamond and 60 wt.% of graphite. The results show that the application of this lubricant reduces friction losses in the engine by 25–30%.

Table 5.1. Characteristics of Running-In of Carburetor Engine when Using Various Lubricants [2]

Characteristic	Industrial oil	Lubricating composition
Number of revolutions of engine crankshaft after switch-off of ignition during run-out after assembling (averaged over three engines)	25–26	25–26
Number of revolutions of engine crankshaft after switch-off of ignition during run-out after running-in at n = 1800 rpm. for 20 min (averaged over three engines)	31–32	47–48
Relative augment of number of revolutions of engine crankshaft during run-out after running-in	15–20	45–50

The tests of industrial oil with 0.1 wt.% of cluster carbon (contains 30 wt.% of diamond and 70 wt.% of graphite) in the lubricating system of high-precision metal-working machines have shown a possibility of its potential application for these purposes (Table 5.2).

Table 5.2. Results of Testing of Modified Oil in Lubricating System of High-Precision Metal-Working Machines [2]

Machine type	Observed effect
Coordinate miller MANO, Germany – lubrication of guides	Rated positional accuracy less than 5 μm
Universal vertical drilling-milling machine BOKO, Germany – lubrication of guides	Cease of oil leakage, formation of an efficient lubricating film, provision of positional accuracy less than 5 μm
Accurate grinding machine Magerle, Switzerland	Machine accuracy retains and oil consumption decreases three times
Coordinate grinding machine Mitsui Seiki, Japan	Decrease in deflections during feed of grinding head from 5 down to 2–3 μm
Boring machine Cross, Germany	Oil consumption reduction four times when retaining rated characteristics
Circular grinder Toyda, Japan	Replacement of oil Vacuoline 1409 by industrial oil with 0.1 wt.% of cluster carbon. Machine accuracy retains at level of 2 μm

The use of the lubricant with the UDDG during bench tests of a diesel engine with gas-turbine supercharge having a power of 130 kW reduces fuel consumption by 2.0–2.5 g/kW·h within the whole range of the rotational speed [1].

The application of the grease containing 0.2 wt.% of cluster carbon (contains 40 wt.% of diamond and 60 wt.% of graphite) to lubricate punches and dies for deep-drawing of cartridges from stainless steel reduced the drawing force by 50–80% and improved article quality compared to the use of the grease containing 10 wt.% of natural graphite with an average particle size of 5 μm [1].

The lubricant developed by the authors is used to lubricate machines in both running-in and operation [1]. It is also applied as an additive to greases for lubrication of sliding bearings, guides, sliders, ball joints, chain drives, and metal-working tools, e.g., molds. It can be used to modify transmission oils for heavy-duty units and motor oils for combustion engines and other mechanisms.

The authors of [4] propose a lubricant containing, in addition to detonation synthesis cluster carbon in an amount of up to 8 wt.%, a nonionic surfactant in an amount of 0.01–8.0 wt.%. The lubricant possesses improved antifriction and antiwear properties owing to variations in its disperse characteristics and the correct proportion between the diamond and graphite modifications of the carbon additive. The carbon additive provides boundary lubrication during unit operation due to the formation of antifriction layers on friction surfaces from ultradispersed diamond and graphite particles. During the fabrication of the lubricant ultradispersed diamond particles, which possess a high adsorptivity, adsorb a film of the surfactant contained in the mineral oil. The protective film formed on particle surfaces provides the stable dispersity of particles and gives them anticoagulation properties. Besides, the particles covered with the protective film serve to form a stable boundary antifriction film. When the lubricant enters the gap between the rubbing surfaces, ultradispersed diamond and graphite particles contained in the carbon additive form additional interlayers, which have a low friction coefficient and a high strength under heavy loads. The lubricant containing surfactants possesses quite good sedimentation stability.

Table 5.3 presents the characteristics and recommended fields of the application of the lubricants based on the plasticized oil [4].

Table 5.3. Tribological Characteristics of Plasticized Oils with Ultradispersed Diamond-Containing Additives

Composition of carbon additive and its concentration in oil, wt.%	Friction coefficient under pressure of 200 kg/cm^2	Wear rate $\times 10^{-6}$, kg/cm	Critical load, N	Seizure load, N	Limiting working temperature, K	Field of application
UDD (4–12 nm) – 20 Graphite (20–100 nm) – 80 Concentration in oil – 0.6	0.010	11.2	2900	4136	460	Loaded axial joints, journal bearings
UDD (6–8 nm) – 30 Graphite (40–80 nm) – 70 Concentration in oil – 0.3	0.025	11.6	2733	4018	440	Lubrication of bearings, chains, and transmissions
UDD (6–8 nm) – 30 Graphite (40–80 nm) – 70 Concentration in oil – 1.0	0.032	14.3	2273	3900	436	High-temperature units operating under heavy pressures
Original oil (without UDDG) plasticized to viscosity of 200 cSt	0.080	60.0	1274	2303	380	

Lubricants of various compositions (grease LT8102: the base is mineral oils, the thickener is complex lithium soap, and the additives are aluminum and copper powder and graphite; assembling spread LT8155: the base is mineral oil with molybdenum disulfide and the thickener is mineral wax; the antifriction coating LT8191: a solid lubricant based on molybdenum disulfide and binding polyepoxy ether) modified by the ultradispersed diamond-graphite charge UDDG are applied in heavy-duty sliding units of heat-and-power plants in the Republic of Belarus. The developed lubricants are used in both restored units containing members made of macroheterogeneous composites (a support of a turbine bearing cage, bearings of the steam-distribution system) and non-restored key joints of turbines in steel–steel and steel – cast iron friction pairs [5–9].

The development of the production technology of diamond-containing additives promotes their wider practical application in lubricants. It involves the stages of the preparation of the diamond-graphite charge and oil, mixing, the tribochemical treatment and conditioning of oil suspensions as well as works on the optimization of treatment conditions [2]. The technological specifications and the process flowchart of the fabrication of modified greases have been developed. Original equipment and devices for manufacture of greases with a set of diamond-containing additives have been produced [3]. The innovation association "Akademtekhnopark" of the National Academy of Sciences of Belarus organized the production of greases with nano-sized additives in an amount of 50 tons per year.

The complex lithium grease ITMOL-150N with a set of diamond-containing additives is a new lubricant with a prolonged life intended for application in heavy-duty friction units. It contains a cross-linked frame from short-fiber filaments possessing better oil-retaining properties that explains its higher carrying capacity (1.6–3.3 times). The grease has the working temperature range widened by 30–40 °C and the life prolonged 1.4–1.7 times. It is intended for lubrication of friction units that operate under heavy loads and at high vibrations (quarry equipment, railway transport, building and road-making machines, and agricultural machinery) as well as at elevated temperatures (rolling mills, forge-and-press machinery, heat-and-power machinery, furnace ventilators, systems of extract and input ventilation, and others).

Technical Characteristics of Grease ITMOL-150N with Set of Diamond-Containing Additives

Critical load, N	no less than 1200
Seizure load, N	no less than 5000
Dropping point, °C	no less than 230
Colloidal stability, %	2–3
Pressure in friction unit, MPa	up to 40
Working temperature range, °C	from –45 to +150 (shortly up to 180)

The method of tribomechanical treatment is used to improve the serviceability of mechanisms of relief copying of harvester reapers. The technology of the formation of antifriction wear-resistant coatings on working surfaces of ball joints of reapers involves the preliminary formation of a copper or bronze film 15–20 μm thick followed by its tribomechanical modification at the initial stage of unit operation. One of the possible variants of the preliminary formation of the copper or bronze film is the method of deformation plating by a flexible tool [10–14].

Table 5.4 shows the schematic of the implementation of tribomechanical modification as applied to joints of harvester reapers.

Table 5.4. Schematic of Implementation of Tribomechanical Modification of Friction Surfaces of Harvester Reaper Joints

<table>
<tr><td colspan="3">1. Preparation of the journal surface (flushing and degreasing)</td></tr>
<tr><td colspan="3">2. Deposition of a copper film 15–20 μm thick</td></tr>
<tr><td>Deformation plating</td><td>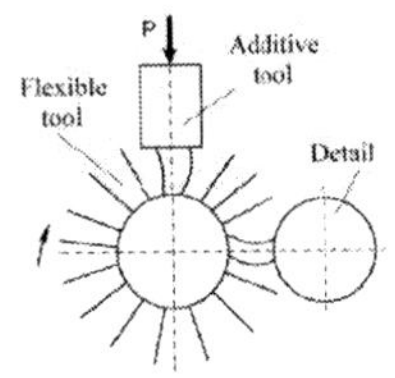
</td><td>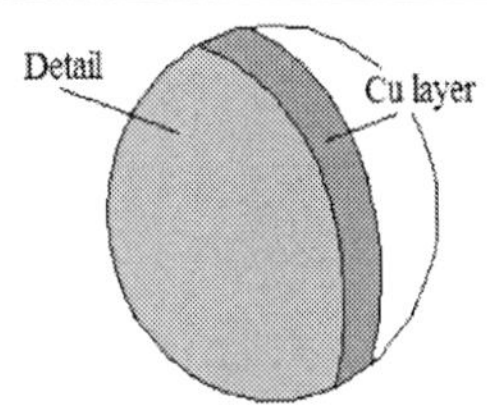
</td></tr>
<tr><td colspan="3">3. Application of a film of the lubricant with nanodiamonds</td></tr>
<tr><td>Lubrication</td><td>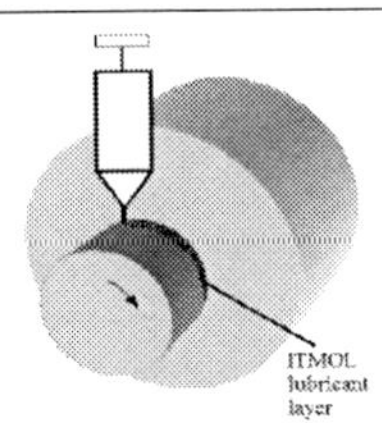
</td><td>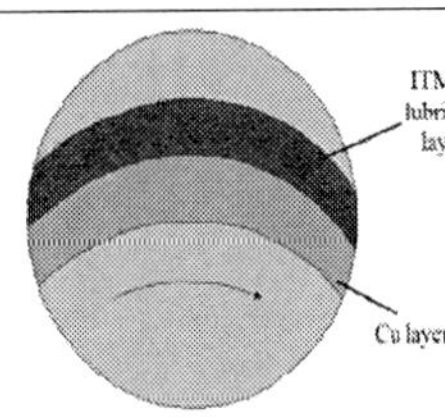
</td></tr>
<tr><td colspan="3">4. Tribomechanical modification during running-in</td></tr>
<tr><td>Tribomechanical modification</td><td>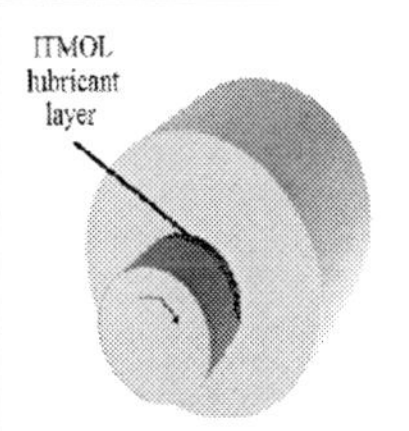
</td><td>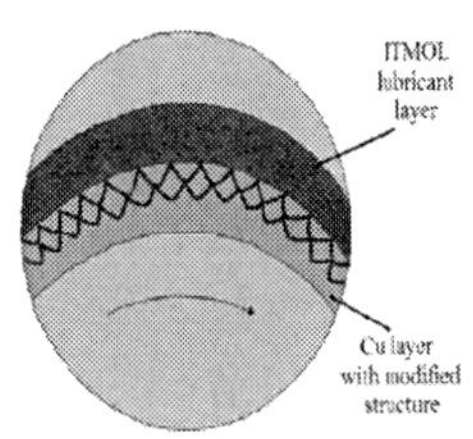
</td></tr>
<tr><td colspan="3">5. Normal operation when using the commercial grease</td></tr>
<tr><td></td><td>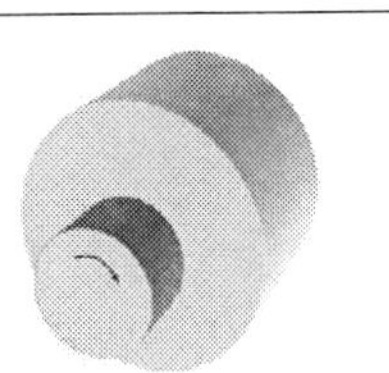</td><td>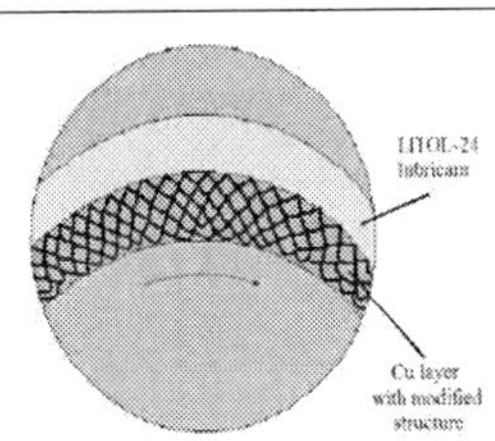
</td></tr>
</table>

The examination of uncoated working surfaces of friction members of mechanisms of relief copying has shown the presence of grooves with the plastically deformed metal, local worn areas, signs of tearing, and seized areas up to 0.6 mm deep. No traces of wear, delamination, and damage were observed on surfaces of the antifriction coatings (figure 5.1).

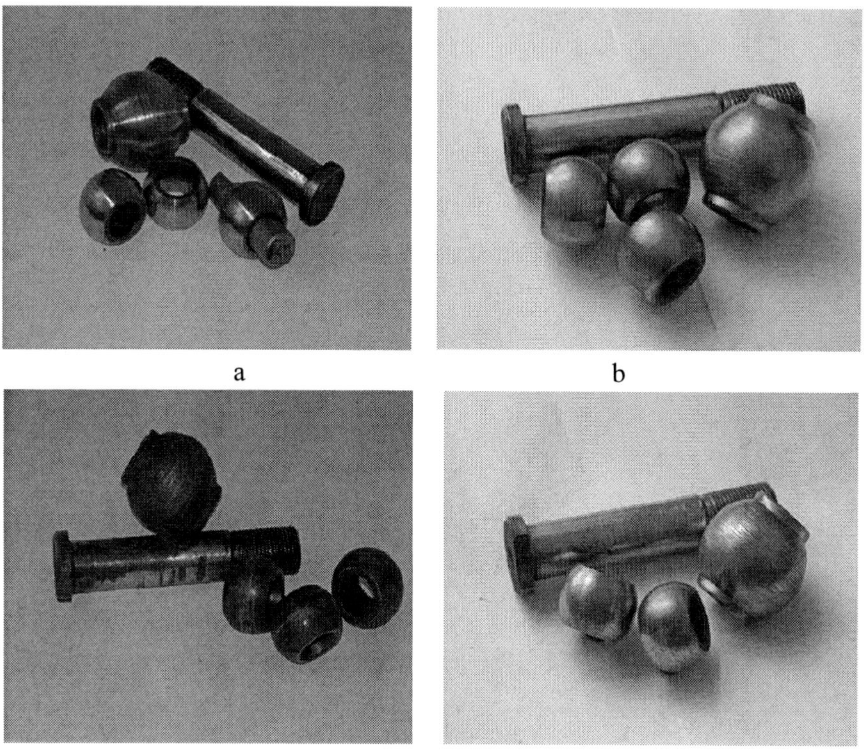

a b c d

Figure 5.1. General view of ball joint members of mechanisms of relief copying of reapers without coatings (*a*, *c*) and with antifriction coating (*b*, *d*): *a*, *b* – before testing; *c*, *d* – after testing.

Results of the comparative tests of hinged-lever mechanisms of relief copying of harvesters have shown that the velocity of the mechanisms comprising modified friction pairs increases 1.5–1.8 times that reduces corn losses during harvesting from 0.78 to 0.52%. In addition, the application of the method of tribomechanical treatment prolonged twice the overhaul period of hinged joints.

The method of tribomechanical treatment is used at the repair and modernization of various friction units of heat-and-power machinery, viz. the support of a turbine bearing cage (figure 5.2, *a*), key joints (figure 5.2, *b*), bearings of the steam-distribution system (figure 5.2, *c*), and the support of a regenerative stove (figure 5.2, *d*).

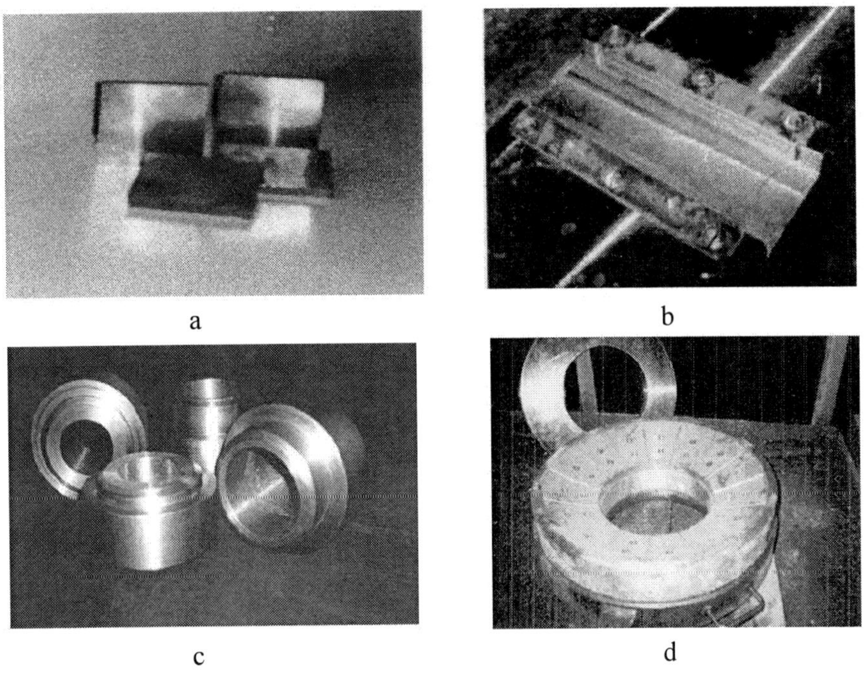

Figure 5.2. Members of sliding units for heat-and-power machinery made of composite materials: *a* – supports of turbine bearing cage; *b* – key joints; *c* – sliding bearings of steam-distribution system; *d* – support of regenerative stove.

The above data allow us to draw a conclusion that lubricants modified by ultradispersed carbon diamond-containing additives hold promise as lubricants for heavy-duty friction units. They are recommended for the lubrication of friction units operating under heavy loads and at high vibrations (quarry equipment, railway transport, building and road-making machines, and agricultural machinery) as well as at elevated temperatures (rolling mills, forge-and-press machinery, and heat-and-power machinery) and others. Their use provides the occurrence of the effect of tribomechanical modification of surfaces that improves the antifriction and antiwear properties of the latter. It prolongs the life of friction units and improves the performance of machines and production equipment.

5.2. Hardening of Gas-Thermal Coatings by Thermal-Force Treatment with Impregnating-Lubricating Compositions Containing Nano-Sized Carbon Components

X-ray structure analysis and X-ray spectrum microanalysis have shown that, after tribomechanical modification with a lubricant comprising diamond nanoparticles, gas-thermal coatings contain an increased amount of carbon. The carbon content in gas-flame coatings is twice as much as that in coatings sprayed by hypersonic metallization. This is explained by the filling of the porous structure of coatings with the lubricant containing nanodiamonds. Since the porosity of gas-flame coatings is twice as much as that of metallization coatings, the content of diamond nanoparticles in pores is also twice greater.

It was proposed to use the particles retained in pores for the subsequent hardening of gas-thermal, particularly, gas-flame coatings. Since at the traditional implementation of the spraying of wire materials the maximal hardness of coatings of various steels is low (35–50 HRC), it is necessary to apply further hardening techniques.

Nowadays a number of methods for increasing the hardness and wear resistance of surface layers of metallic materials are available. However, their applicability to the solution of the problem is restricted. In particular, laser and electron beam treatment result in the inhomogeneous macrostructure of surfaces and do not provide their chemical modification. High-energy ion implantation methods increase considerably the coating cost. Thermochemical treatment is characterized by a long duration of the process and is not suitable for treating large articles. Therefore, the development of an effective method for improving the strength and wear resistance of gas-thermal coatings is an urgent task today.

The direct passing of electric current (electrocontact treatment) seems to be the most efficient and profitable way of providing high hardness, wear resistance, and other surface properties of gas-thermal coatings of iron-based alloys. It was assumed that high-rate electrocontact heating combined with activating pressure facilitates the diffusion of carbon nanocomponents into the coating structure with the formation of atomic and molecular bonds with the basic coating components. The use of a combination of spraying technologies, tribomechanical modification, and subsequent electrocontact (electric-force) treatment opens wide possibilities of developing composite coatings with specific properties. Such combined technology requires no additional expensive equipment and stages that reduces the cost of the hardening and recovery processes.

Gas-thermal coatings of wire steels belonging to martensite (steel 40Kh13) and ferrite (steel SV-08) classes were selected as objects of the study. The conditions of tribomechanical modification were presented in Sections 3.3 and 4.4.

Figure 5.3. General view of setup for thermal-force treatment of coatings.

Conditions of thermal-force treatment were selected using an UEKS-2 setup with a modified control unit RKS-801M (figure 5.3). The current strength in the secondary circuit was measured by a Rogovskii belt and an amperemeter Duffer Associates (USA). Current and voltage parameters were recorded by a light-beam oscillograph. Oscillograms were processed in a PC followed by plotting of changes in the electric resistance of the deposited layer during electric-force treatment. Dry friction tests were carried out using an automated tribometer ATVP.

It has been found that the thermal-force treatment of coatings of steel 40Kh13 after their tribomodification increases considerably the surface layer microhardness (up to H_{μ} = 8400–8500 MPa). The analysis of the results shows (figures 5.4 and 5.5) that the hardened layer extends almost through the total coating depth and its thickness reaches ≈ 0.20–0.25 mm. The phase composition of the modified layer does not practically change compared to the original state. However, slight increase in the content of retained austenite in the layer (up to $V_{\gamma} \approx$ 25–27 vol.%) and in the lattice spacing of austenite (up to $a_{\gamma} \approx$ 0.3618 nm) is found.

Besides, the modified layer demonstrates a somewhat greater physical widening of diffraction lines of the matrix α-phase. An increased content of retained austenite and greater lattice spacings of the α- and γ-phases prove a high carbon content in them resulted from the dissolution of nanodiamonds kept in pores during thermal-force treatment. The increased surface microhardness of the modified coating of steel 40Kh13 hardened by thermal-force treatment can be attributed to a high carbon content in the modified layer and the martensite transformation of retailed austenite during coating grinding. No cracks were observed in the layer modified by thermal-force treatment.

The thermal-force treatment of the modified coating of steel SV-08 (the steel has the following chemical composition, wt.%: C – 0.08, Si – 0.27, Mn – 0.50, Cr – 0.10, Ni – 0.25, Cu – 0.25, and the rest is Fe) increases the microhardness of the surface layer up to ≈ 6000 MPa. No microcracks were detected in the layer. For low-alloy steel Sv-08 the depth of the hardened layer is small and does not exceed 50 μm (figures 5.4 and 5.5). The phase composition of the coating of steel SV-08 modified by thermal-force treatment contains the α-phase, the γ-phase (austenite), oxide FeO as well as Fe_3C and carbon in small amounts. The appearance of such structural components as austenite ($V_{\gamma} \approx$ 5 vol.%), Fe_3C, and carbon in the layer phase composition proves the partial dissolution of nanodiamonds during thermal-force treatment. However, because of a short duration of treatment and poor solubility of carbon in low-alloy steel Sv-08, the layer depth does not exceed 50 μm and cementite particles appear at coating cooling after treatment (figure 5.6).

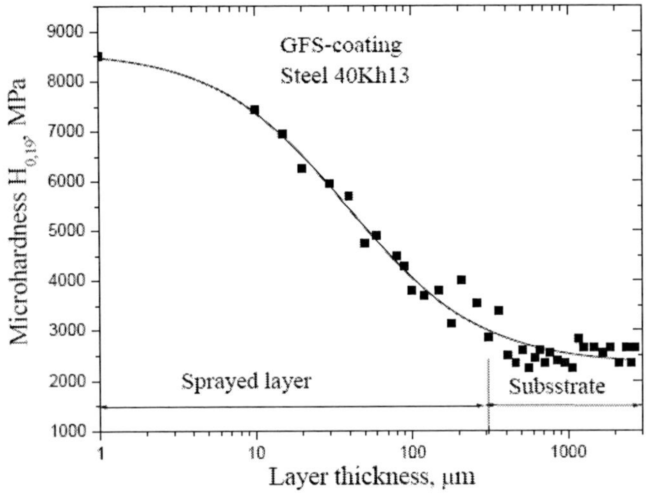

Figure 5.4. Distribution of microhardness in depth of modified coating from steel 40Kh13 hardened by thermal-force treatment.

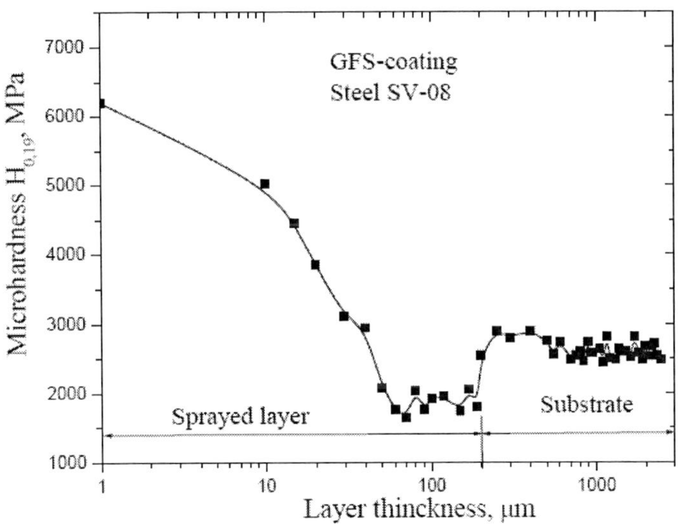

Figure 5.5. Distribution of microhardness in depth of modified coating from steel SV-08 hardened by thermal-force treatment.

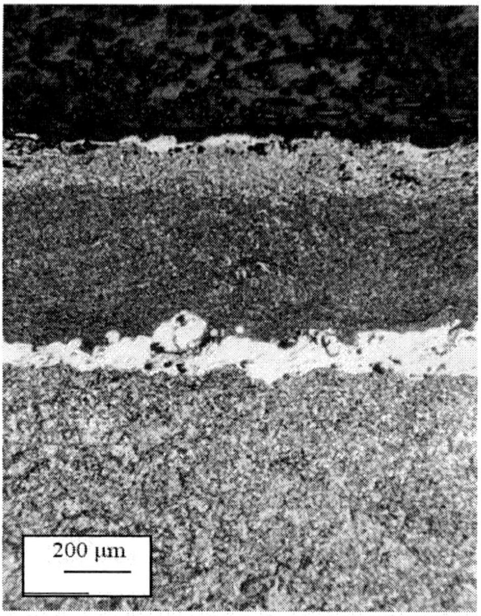

a

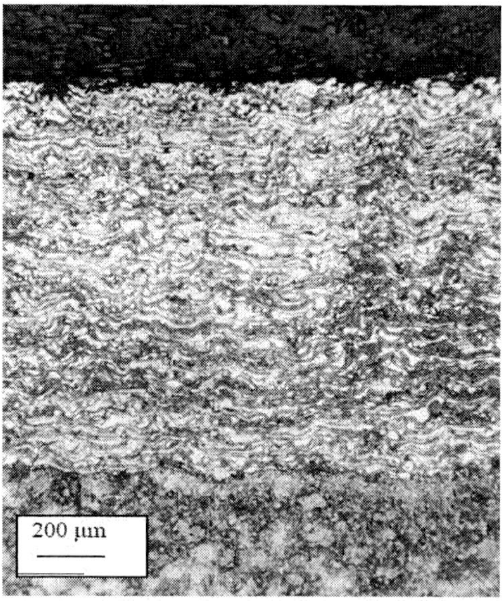

b

Figure 5.6. Microstructure of modified gas-thermal coatings hardened by thermal-force treatment: *a* – steel SV-08; *b* – steel 40Kh13.

Results of tribotests of the gas-thermal coatings under study are presented in figures 5.7–5.10.

Figure 5.7 shows the dependences of the mass wear on the sliding distance for a wire coating of steel 40Kh13 in the original state as well as for the same coatings hardened by thermal-force treatment and tribomodified with subsequent hardening by thermal-force treatment. The values of the mass wear of various coatings differ considerably. In particular, the plots in figure 5.7 show that the original gas-thermal coating has a comparatively low wear resistance; the average wear rate of the layer is I_q = 1.1 mg/m. The friction coefficient is $f \approx 0.7$ (figure 5.8).

The thermal-force treatment of the coating of steel 40Kh13 increases significantly its wear resistance. The wear rate of the treated coating without tribomodification decreases down to I_q = 0.07 mg/m, while that of the tribomodified coating subjected to thermal-force treatment decreases down to I_q = 0.05 mg/m. Thus, thermal-force treatment increases the wear resistance of coatings of steel 40Kh13 from 15 (in case of the nonmodified coating) to 20 times (for the preliminary modified coating).

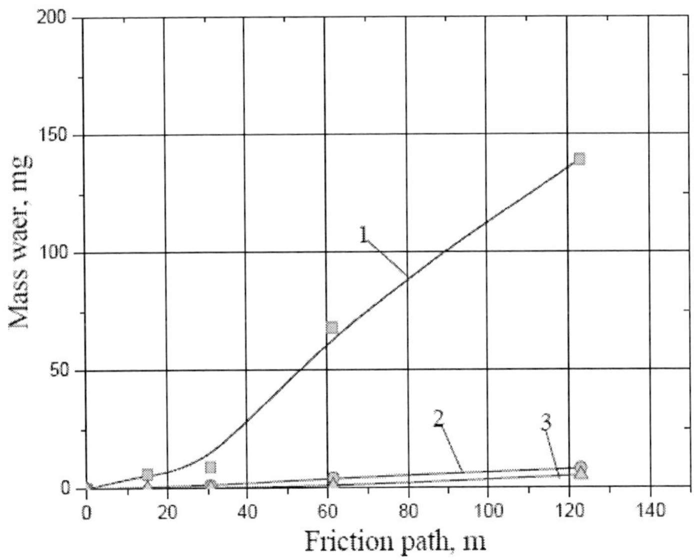

Figure 5.7. Dependence of mass wear on sliding distance for wire coatings from steel 40Kh13 treated under various conditions (dry friction under pressure p_a = 3 MPa): *1* – GFS-coating from steel 40Kh13 (ground); *2* – ditto + thermal-force treatment; *3* – ditto + modification + thermal-force treatment.

Increase in the wear resistance of the modified layer is explained by both the growth of its microhardness and a higher content of retained austenite in it. It should be borne in mind that the presence of metastable austenite in the layer hampers the nucleation and propagation of microcracks that yield wear particles. In addition, the transformation of austenite to martensite during friction also hardens surface layers and induces compressive stresses in them, which inhibit friction failure. The dry friction coefficient of the coatings hardened by thermal-force treatment increases up to f = 0.9–1.0. The high friction coefficient proves the rise of the activation energy of the failure process of hardened layers of the coatings compared to the nonmodified layer.

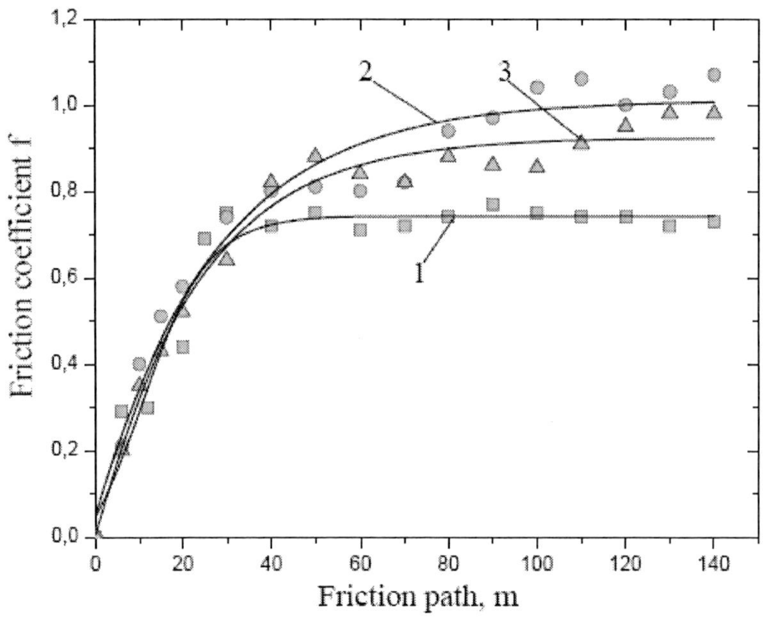

Figure 5.8. Dependence of friction coefficient on sliding distance for wire coatings from steel 40Kh13 treated under various conditions (dry friction under pressure p_a = 3 MPa): *1* – GFS-coating from steel 40Kh13 (ground); *2* – ditto + thermal-force treatment; *3* – ditto + modification by UDDG + thermal-force treatment.

Coatings from low-alloy steel SV-08 have a much lower wear resistance (figure 5.9). In particular, the wear rate of the sprayed coating in the original (untreated) state is $I_q \approx 7$ mg/m. The thermal-force treatment of the coating reduces the mass wear at the initial stages of testing two times.

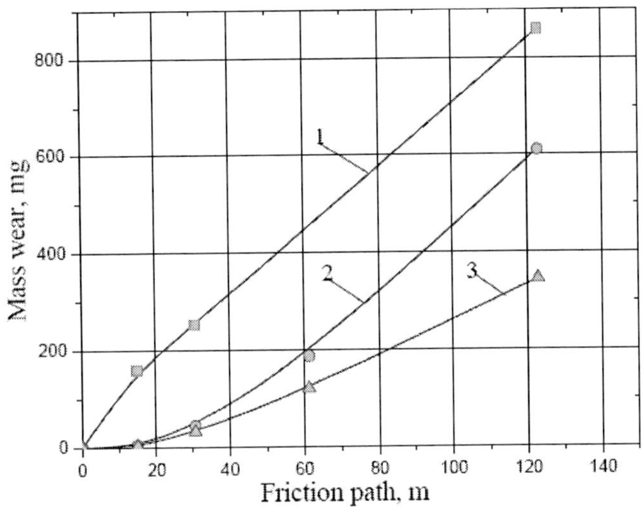

Figure 5.9. Dependence of mass wear on sliding distance for wire coatings from steel Sv-08 treated under various conditions (dry friction under pressure $p_a = 3$ MPa): *1* – GFS-coating from steel SV-08 (ground); *2* – ditto + thermal-force treatment; *3* – ditto + modification by UDDG + thermal-force treatment.

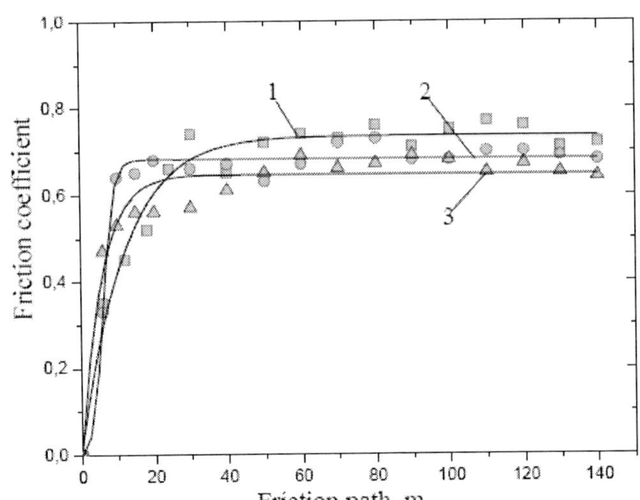

Figure 5.10. Dependence of friction coefficient on sliding distance for wire coatings from steel Sv-08 treated under various conditions (dry friction under pressure $p_a = 3$ MPa): *1* – GFS-coating from steel SV-08 (ground); *2* – ditto + thermal-force treatment; *3* – ditto + modification by UDDG + thermal-force treatment.

After passing a sliding distance of 50 m, the wear rate of the coating from steel SV-08 subjected to thermal-force treatment reaches the value typical for the untreated coating. This is confirmed by the slope angle of curves *1* and *2* in figure 5.9. The friction coefficient of the coatings is $f = 0.6–0.7$ (figure 5.10). The coating from steel SV-08 tribomodified and subjected to thermal-force treatment has a higher wear resistance, but the coating mass wear at the late stages of testing also increases considerably.

Thus, thermal-force treatment has the most pronounced hardening effect on modified coatings from high-alloy martensite steels.

References

[1] Patent PCT/ SU, 87/00249 (SU)

[2] Dolmatov V.Yu. Ultradispersed Diamonds Produced by Detonation Synthesis. Production, Properties, and Applications; St Petersburg State Polytechnic University: St Petersburg, RU, 2003, 344 p.

[3] Ivakhnik, A.V., Pinchuk, T.I., Zhornik, V.I. Study of Structure Formation of Dispersed Phase of Grease by Scanning Electron Microscopy; Chemical Reagents and Processes of Small-Tonnage Chemistry; Belarus State Technological University: Minsk, BY, 2005, pp 43-44.

[4] Patent PCT/SU WO 93/01261 (SU)

[5] Zhornik, V.I., Kalinichenko, A.S. et al. Recommendations on Repair and Reconstruction of Heavy-Duty Sliding Units Using Composite Materials; Institute of Technical Cybernetics of National Academy of Sciences of Belarus: Minsk, BY, 2000; 88 p.

[6] Kobzar, Yu.V., Vykhota, S.O. et al. *Power Engineering. 2001*, N 4, 73-78.

[7] Zhornik, V.I., Kukareko, V.A. Transactions of Polotsk State University. *Applied Sciences.* 2005, N 6, 150-156.

[8] Kalinichenko, A.S., Zhornik, V.I. et al. Application of Composite Materials and Diamond-Containing Lubricating Coatings in Heavy-Duty Sliding Units; Theoretical and Technological Basis for Hardening and Recovering of Machine-Building Articles; ed. by S.A. Astapchik and P.A. Vityaz; Tekhnoprint: Minsk, BY, 2001; pp 171-175.

[9] Zhornik, V.I., Kalinichenko, A.S. et al. Repair, Recovering, and Modernization. 2002, N 9, 28-32.

[10] Levantsevich, M.A. Transactions of Polotsk State University. *Applied Sciences;* 2003, Vol 2, N 4, 53-55.

[11] Polzer, G. New Results of Brass-Plating of Steel and Cast Iron Friction Surfaces; Durability of Friction Members of Machines; Mashinostroenie: Moscow, SU, 1987; Vol 2, pp. 81-85.

[12] Belevskii, L.S. Plastic Deformation of Surface Layer and Coating Formation at Deposition by Flexible Tool; Lyceum of Russian Academy of Sciences: Magnitogorsk, RU, 1996; 231 p.

[13] Antsupov, V.P. Technological Basis of Production of Bimetallic Articles Using Plating by Flexible Tool: Magnitogorsk Mining and Smelting Academy Named after G. I. Nosov: Magnitogorsk, RU, 1997, 43 p.

[14] Levantsevich, M.A., Maksimchenko, N.N. et al. Transactions of National Academy of Sciences of Belarus. *Physical-Technical Series.* 2005, N 1, 67-72.

Conclusions

Nowadays the application of methods of tribotechnology is one of the most effective ways of solving the problem of the improvement of the reliability and durability of friction members. In particular, the implementation of conditions of selective transfer at friction (the effect of wearless friction), the use of methods of the non-dismantling recover of friction pairs during their operation with the application of surface remetallizers, cermets, metal conditioners etc. allow one to increase considerably the serviceability of friction units and prolong the life of machines. During recent years, methods of tribotechnology were extended by a new technology of the modification of friction surfaces and the improvement of their properties. It involves nanodiamonds as a basic tool affecting tribosystems. Nanodiamonds belong to solid friction modifiers. Owing to a high surface energy they possess good structure-forming properties, while their high hardness intensifies the plastic deformation of friction surfaces. Nanodiamonds affect the tribosystem members in two ways. First, they change the structure and properties of a lubricant and increase the carrying capacity of a lubricating film. Second, during friction they facilitate structural transformations in surface layers of the contacting solids. Such tribomodification results in a higher hardness and wear resistance of friction surfaces.

The present work illustrates some possibilities of using nanodiamonds as an additive to oils and greases. The efficiency of their application has been confirmed for friction pairs made of metallic, metal-polymer, and polymeric materials as well as gas-thermal coatings. Diamond-containing lubricants possess high tribological characteristics, shorten considerably the running-in of friction units and form wear-resistant structures in surface layers during initial stages of the operation of friction units. The lubricants modified by nanodiamonds were widely tested in friction units of metal-working machinery, combustion engines, heat-and-power plants, and quarry and agricultural machinery etc. The experience of their application has shown that the modification of lubricants by nanodiamonds allows one to widen substantially the ranges of working temperatures and loads, increase the wear resistance of friction pairs and reduce operating cost.

However, despite doubtless progress in the study of the efficiency of the use of nanodiamonds as a tribological material, a number of problems have not yet been solved that restricts the field of the rational application of nanodiamonds in tribology. In particular, to set the boundaries of this field and to expand them a way should be passed involving the investigation of tribological processes in tribosystems containing nanodiamonds, the development of new colloidal-stable lubricants with nanodiamonds and the testing of the new materials.

The presented materials are a section of this path.

Index

D

E

R

S